Table of Contents

Section One: Multiplication..5

   Lattice Multipy 3 Digits by 2 Digits-----------------------------6

   Lattice Multipy 3 Digits by 3 Digits----------------------------16

Section Two: Division...31

   Divide 3 Digits by 1 Digit-------------------------------------32

   Divide 3 Digits by 2 Digits-------------------------------------40

Section Three: Fractions...55

   Adding Fractions--56

   Adding Mixed Numbers-----------------------------------62

   Subtracting Fractions------------------------------------71

   Subtracting Mixed Numbers----------------------------77

Section Four: Decimals...86

   Adding Decimals--87

   Subtracting Decimals-------------------------------------93

   Multiplying Decimals-------------------------------------99

Solutions..105

   Section One---106

   Section Two---120

   Section Three---127

   Section Four--130

School Days Publishing is your source for high-quality, aesthetic and fun
- notebooks
- workbooks
- activity books

Visit us at **www.amazon.com/author/schooldays**

If you find value in this book, please rate and review it. You'll be helping us to develop and publish more great books.

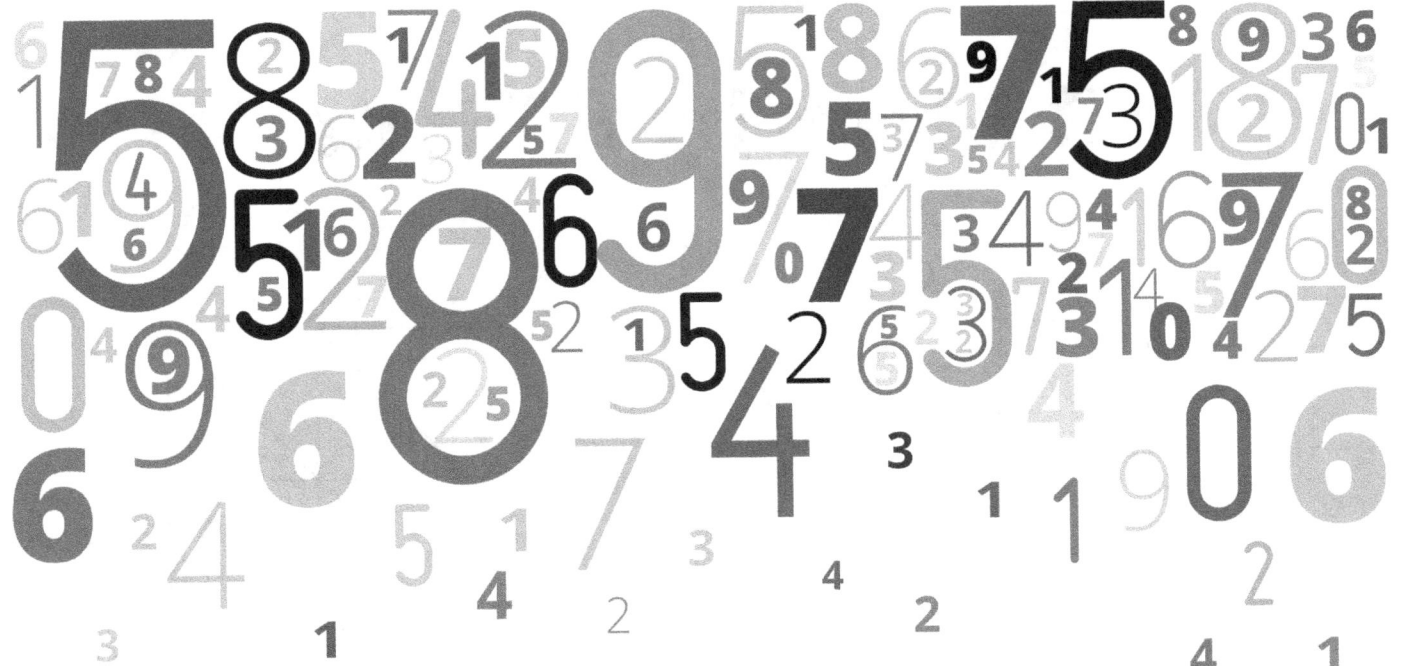

Section One
Multiplication

3 Digits by 2 Digits

①   4   4   0   ×

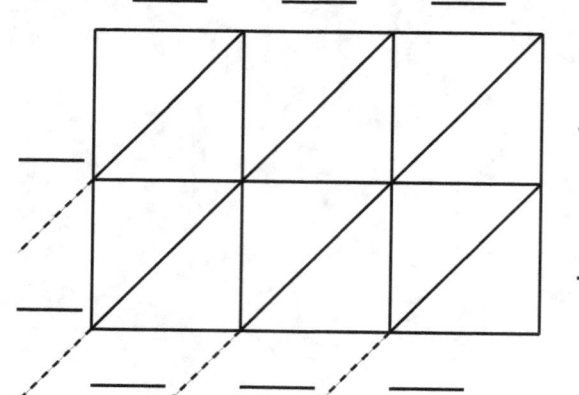

9

4

440 × 94 = _____

②   8   6   1   ×

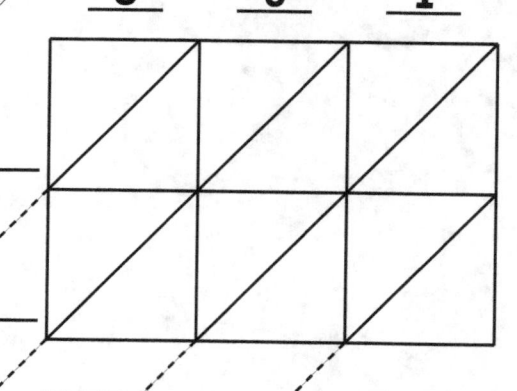

3

4

861 × 34 = _____

③   2   6   3   ×

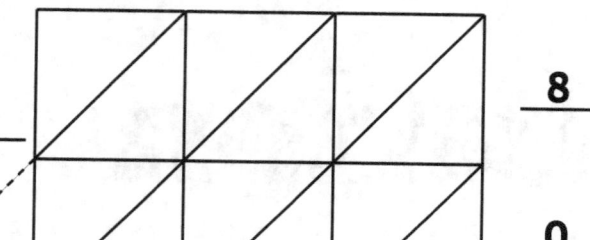

8

0

263 × 80 = _____

④   7   4   5   ×

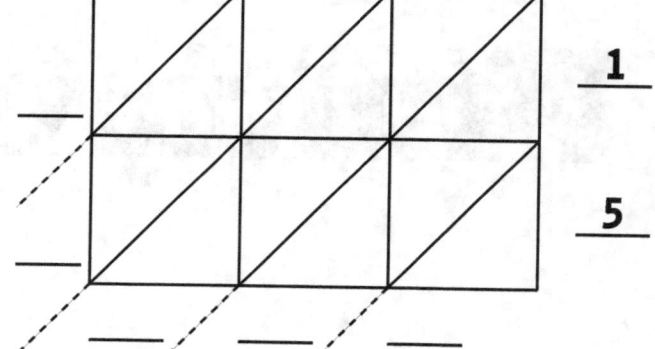

1

5

745 × 15 = _____

⑤   1   2   6   ×

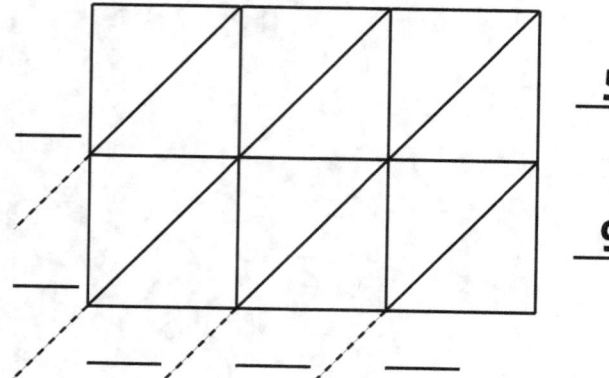

5

9

126 × 59 = _____

⑥   8   6   8   ×

2

0

868 × 20 = _____

3 Digits by 2 Digits

①   3   5   8   ×

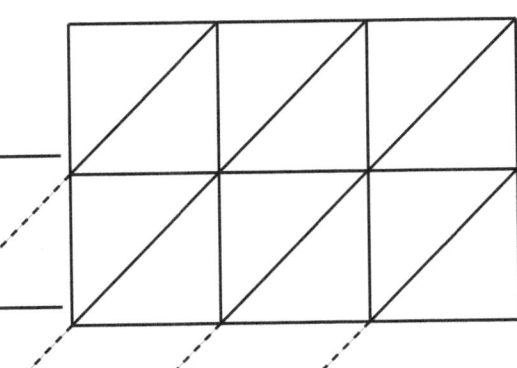

8

3

②   6   6   7

5

7

$358 \times 83 =$ _____

$667 \times 57 =$ _____

③   2   7   7   ×

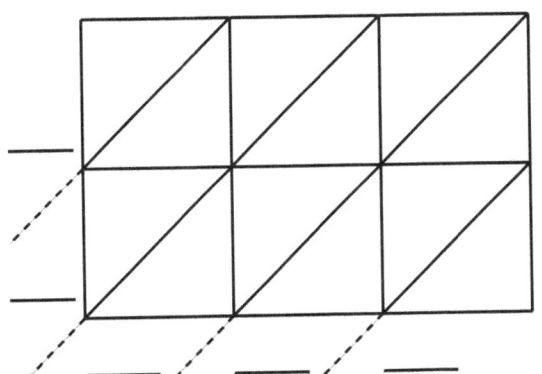

2

8

④   9   9   4   ×

6

1

$277 \times 28 =$ _____

$994 \times 61 =$ _____

⑤   8   1   5   ×

6

5

⑥   1   7   0   ×

6

2

$815 \times 65 =$ _____

$170 \times 62 =$ _____

3 Digits by 2 Digits

① <u>3</u>   <u>5</u>   <u>9</u>   ×

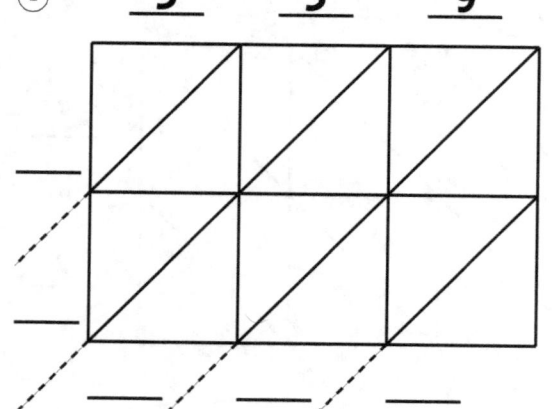

2

9

359 × 29 = _____

② <u>8</u>   <u>3</u>   <u>6</u>   ×

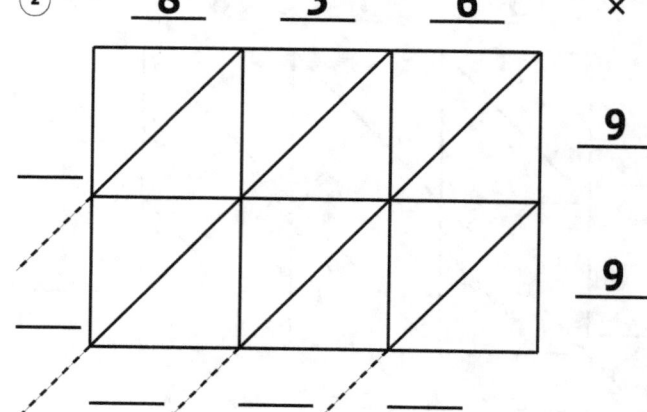

9

9

836 × 99 = _____

③ <u>6</u>   <u>9</u>   <u>5</u>   ×

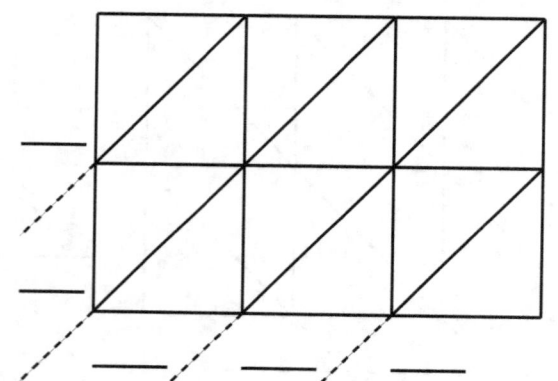

8

8

695 × 88 = _____

④ <u>6</u>   <u>0</u>   <u>2</u>   ×

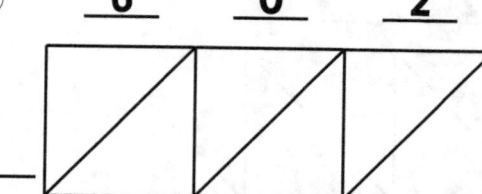

5

3

602 × 53 = _____

⑤ <u>1</u>   <u>6</u>   <u>7</u>   ×

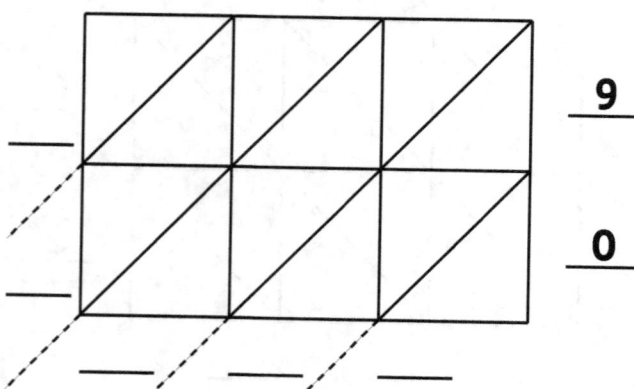

9

0

167 × 90 = _____

⑥ <u>8</u>   <u>1</u>   <u>7</u>   ×

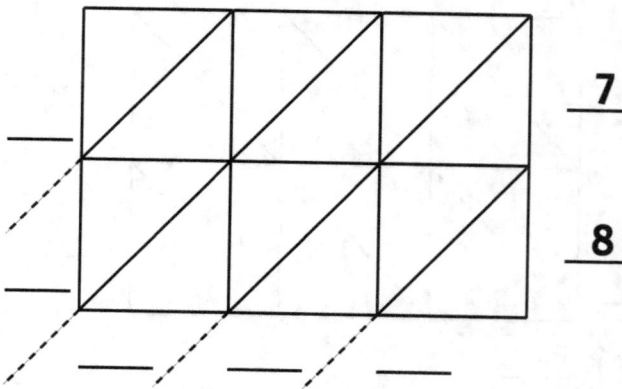

7

8

817 × 78 = _____

3 Digits by 2 Digits

① <u>1</u> <u>4</u> <u>4</u> ×
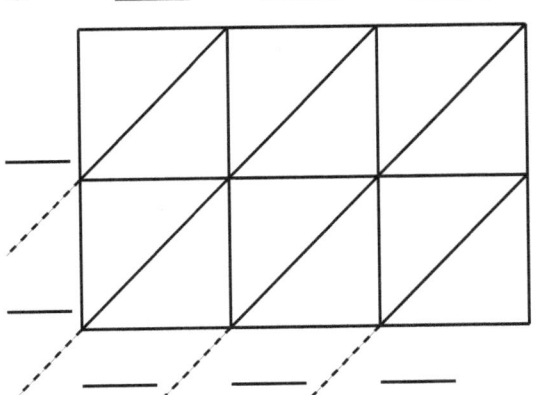
9

5

144 × 95 = _____

② <u>7</u> <u>4</u> <u>2</u> ×
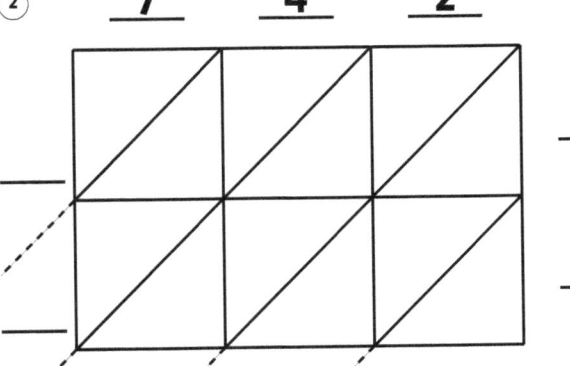
8

5

742 × 85 = _____

③ <u>9</u> <u>4</u> <u>7</u> ×
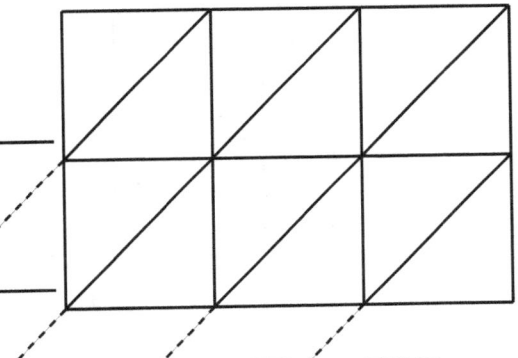
5

2

947 × 52 = _____

④ <u>8</u> <u>6</u> <u>9</u> ×
2

0

869 × 20 = _____

⑤ <u>5</u> <u>4</u> <u>2</u> ×
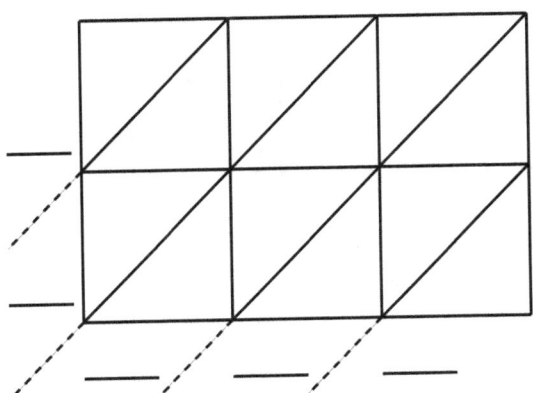
4

8

542 × 48 = _____

⑥ <u>7</u> <u>8</u> <u>1</u> ×
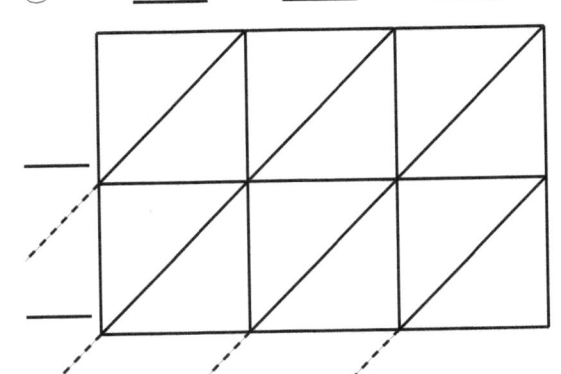
5

1

781 × 51 = _____

9

3 Digits by 2 Digits

①   __8__   __3__   __8__   ×

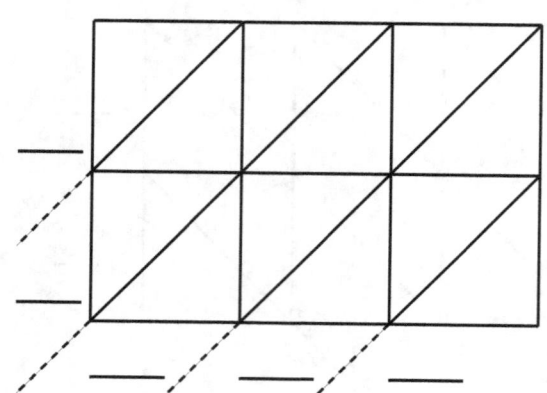

      __1__

      __8__

838 × 18 = _____

②   __1__   __2__   __9__   ×

      __2__

      __9__

129 × 29 = _____

③   __6__   __6__   __0__   ×

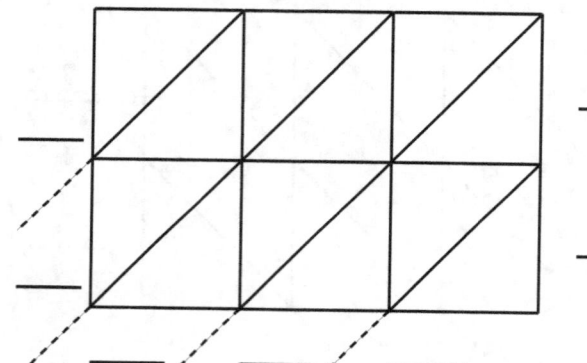

      __5__

      __1__

660 × 51 = _____

④   __7__   __5__   __1__   ×

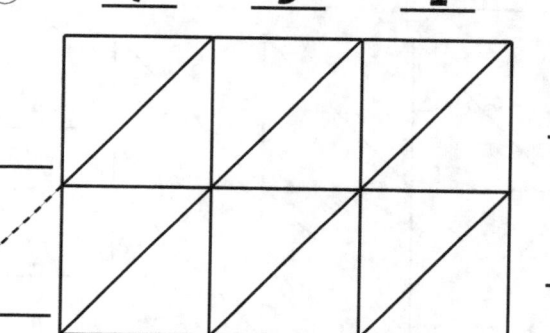

      __1__

      __6__

751 × 16 = _____

⑤   __1__   __9__   __3__   ×

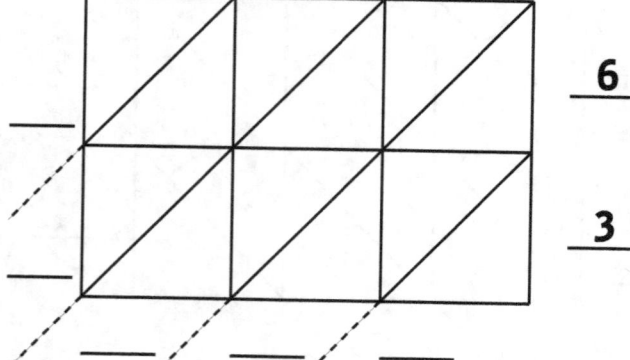

      __6__

      __3__

193 × 63 = _____

⑥   __7__   __6__   __4__   ×

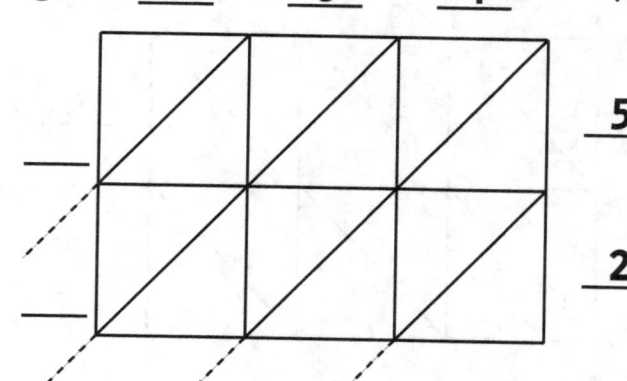

      __5__

      __2__

764 × 52 = _____

3 Digits by 2 Digits

①   5   7   6   ×

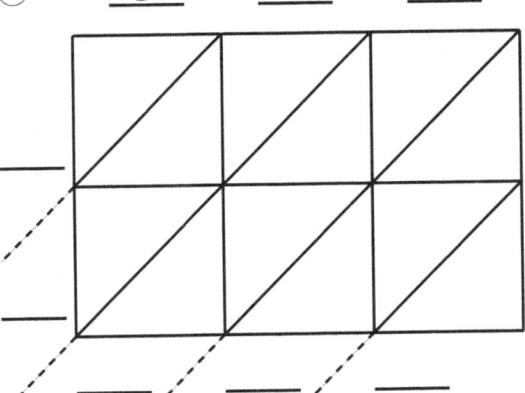

8

1

②   4   8   4   ×

3

3

576 × 81 = _____

484 × 33 = _____

③   2   4   2   ×

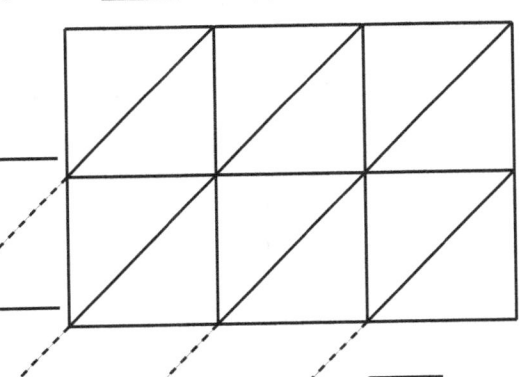

1

2

④   3   0   6   ×

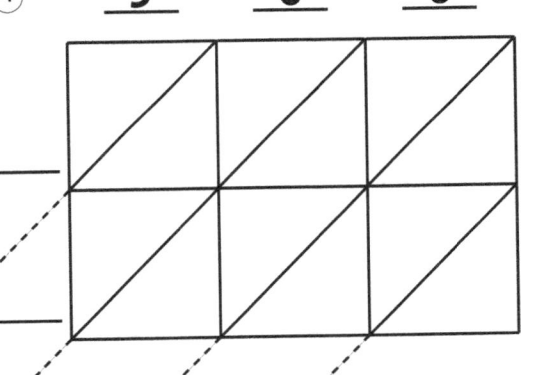
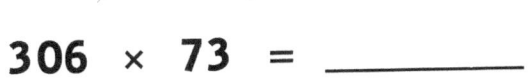

7

3

242 × 12 = _____

306 × 73 = _____

⑤   2   1   1   ×

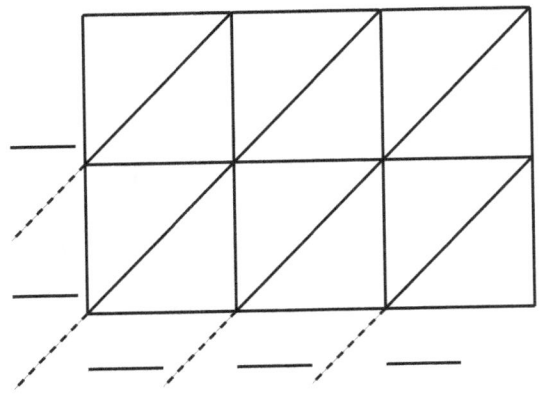

2

2

⑥   5   1   5   ×

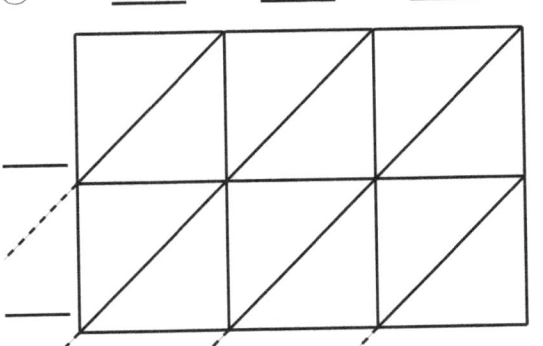

2

4

211 × 22 = _____

515 × 24 = _____

3 Digits by 2 Digits

① ___8___ ___5___ ___4___ ×

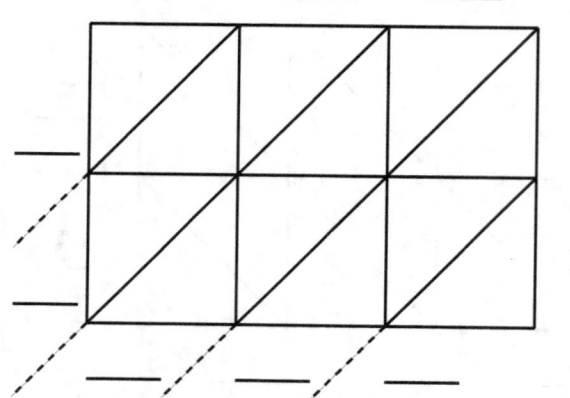

1

5

854 × 15 = _____

② ___6___ ___3___ ___1___ ×

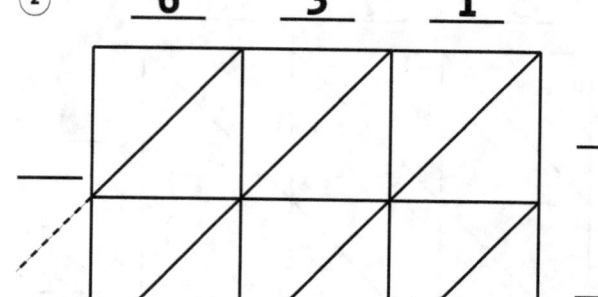

3

7

631 × 37 = _____

③ ___9___ ___0___ ___1___ ×

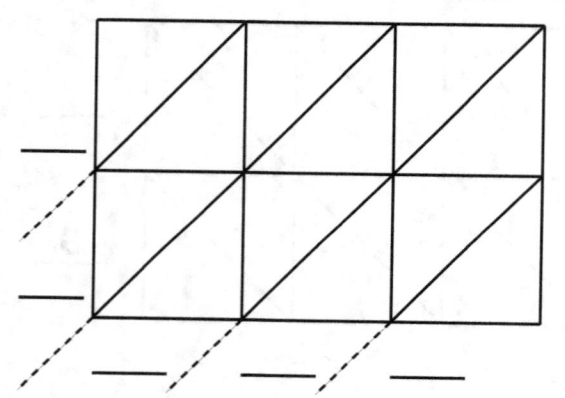

8

4

901 × 84 = _____

④ ___3___ ___8___ ___3___ ×

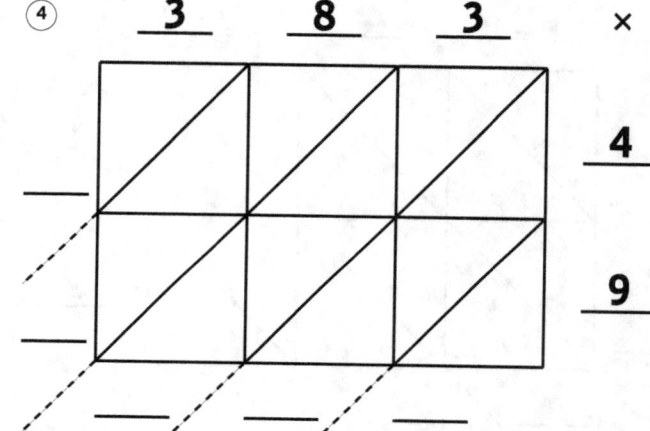

4

9

383 × 49 = _____

⑤ ___7___ ___8___ ___9___ ×

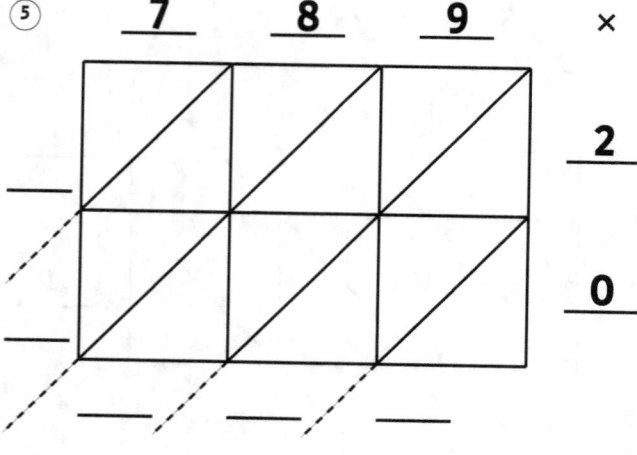

2

0

789 × 20 = _____

⑥ ___3___ ___4___ ___8___ ×

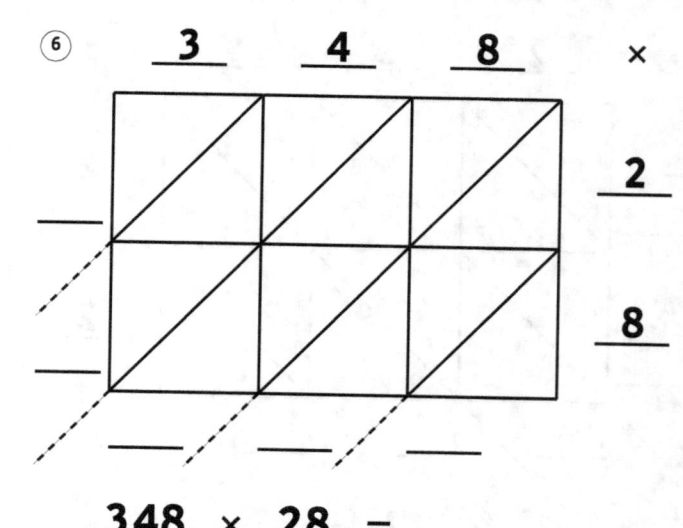

2

8

348 × 28 = _____

12

3 Digits by 2 Digits

1. ___8___ ___0___ ___0___ ×

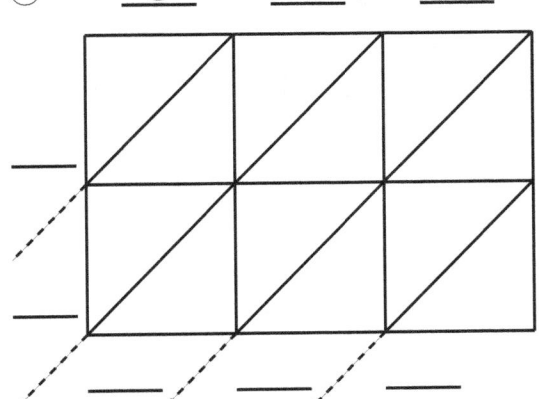

3

5

800 × 35 = _____

2. ___8___ ___0___ ___8___ ×

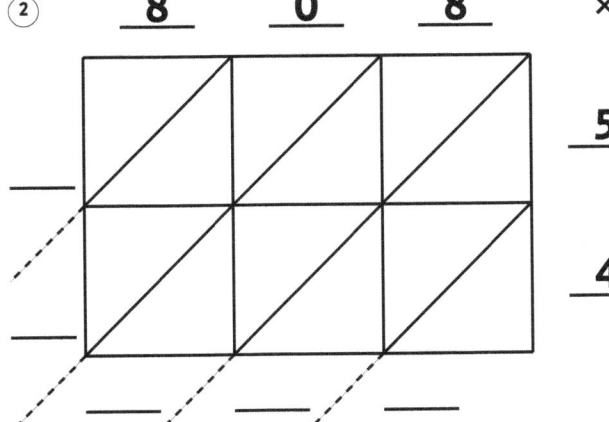

5

4

808 × 54 = _____

3. ___6___ ___3___ ___3___ ×

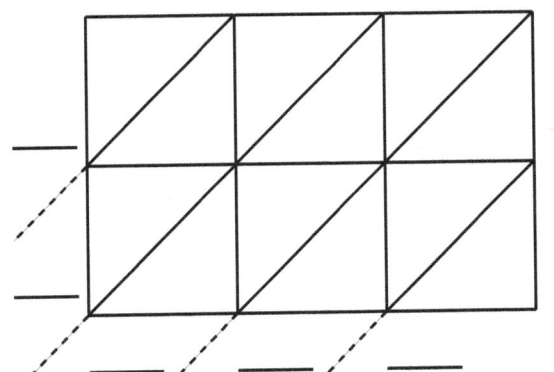

1

5

633 × 15 = _____

4. ___4___ ___2___ ___2___ ×

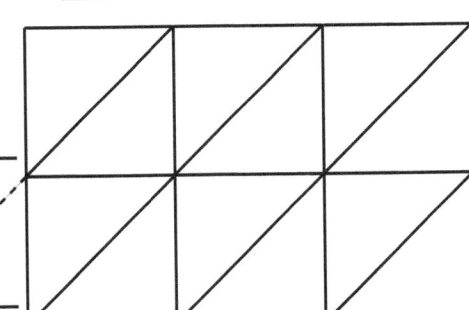

4

8

422 × 48 = _____

5. ___5___ ___4___ ___6___ ×

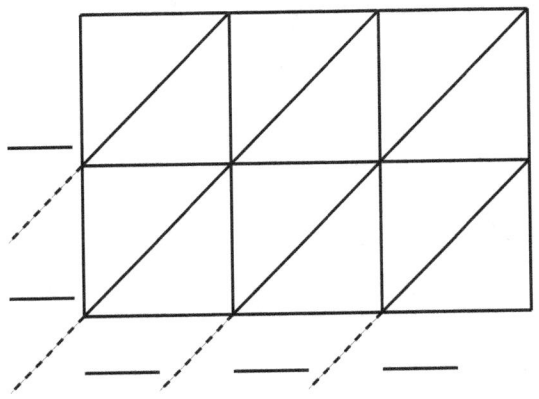

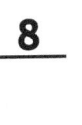

8

6

546 × 86 = _____

6. ___2___ ___9___ ___6___ ×

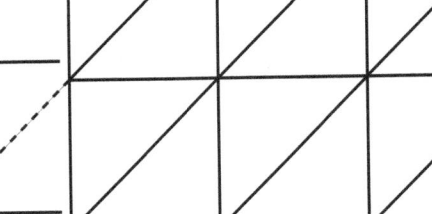

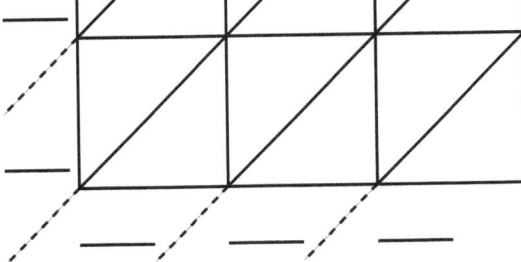

9

1

296 × 91 = _____

3 Digits by 2 Digits

① <u>6</u> <u>5</u> <u>5</u> ×
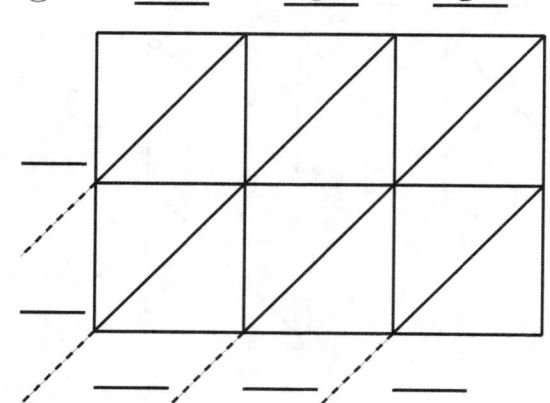
5
0

$655 \times 50 =$ _____

② <u>7</u> <u>8</u> <u>3</u> ×
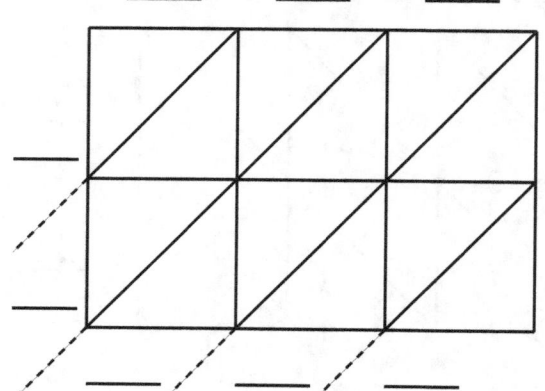
3
2

$783 \times 32 =$ _____

③ <u>9</u> <u>1</u> <u>2</u> ×
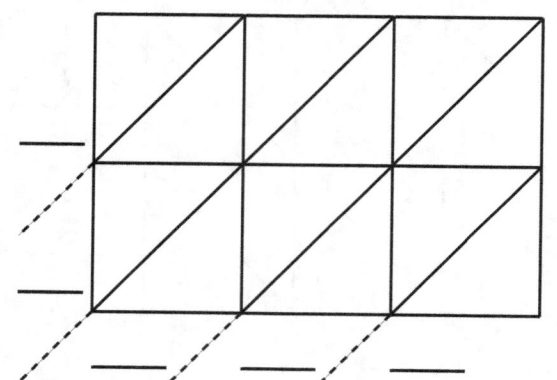
9
3

$912 \times 93 =$ _____

④ <u>8</u> <u>4</u> <u>0</u> ×
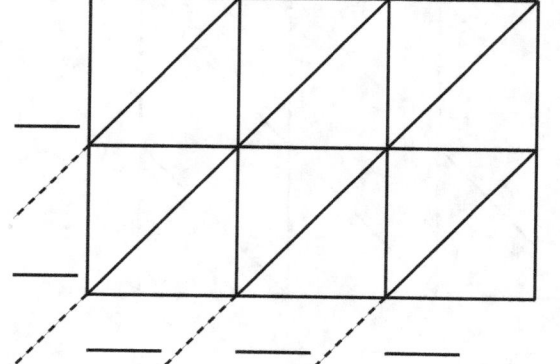
4
0

$840 \times 40 =$ _____

⑤ <u>7</u> <u>1</u> <u>5</u> ×
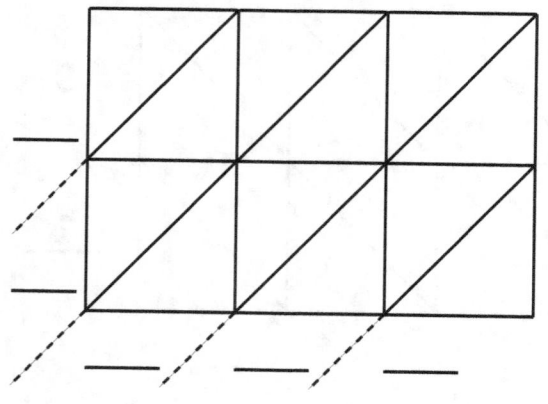
8
4

$715 \times 84 =$ _____

⑥ <u>6</u> <u>2</u> <u>4</u> ×
2
2

$624 \times 22 =$ _____

3 Digits by 2 Digits

(1)   2   1   2   ×

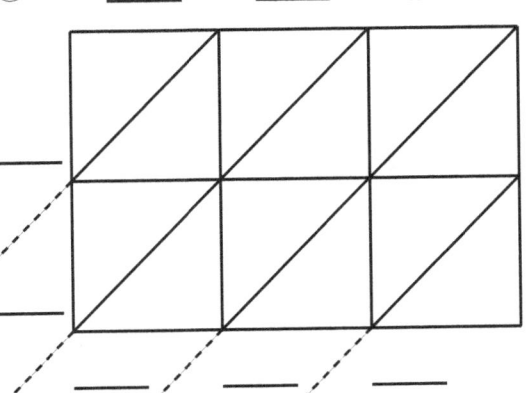

2

4

212 × 24 = _____

(2)   7   1   9   ×

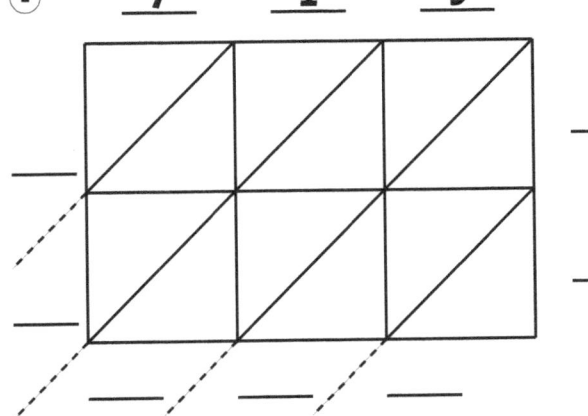

3

3

719 × 33 = _____

(3)   5   8   2   ×

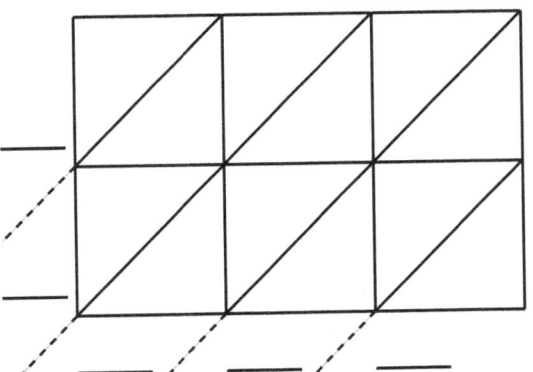

7

5

582 × 75 = _____

(4)   3   5   5   ×

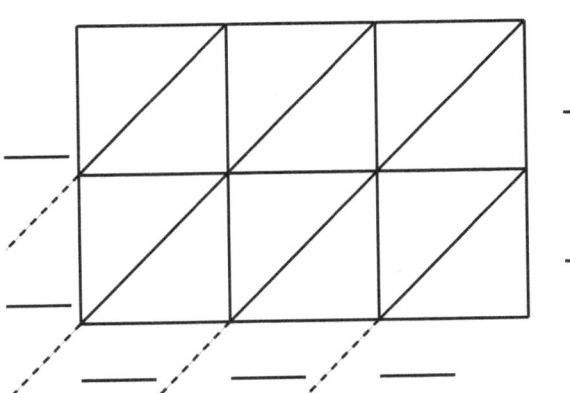

8

3

355 × 83 = _____

(5)   4   4   7   ×

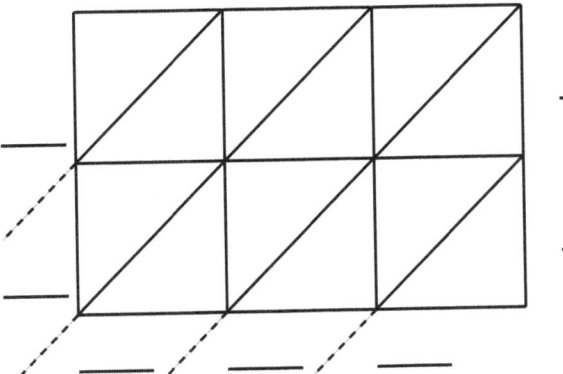

9

1

447 × 91 = _____

(6)   4   5   4   ×

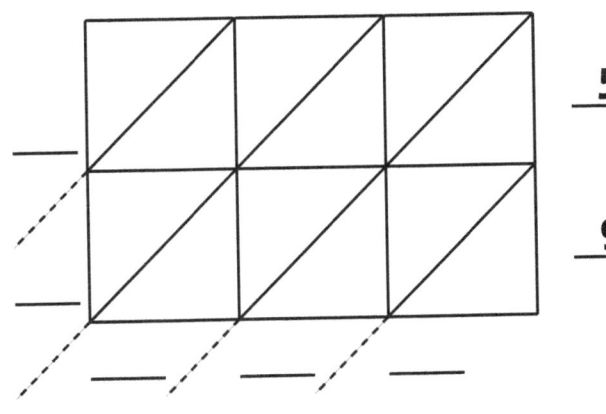

5

9

454 × 59 = _____

15

3 Digits by 3 Digits

①   6   0   0   ×

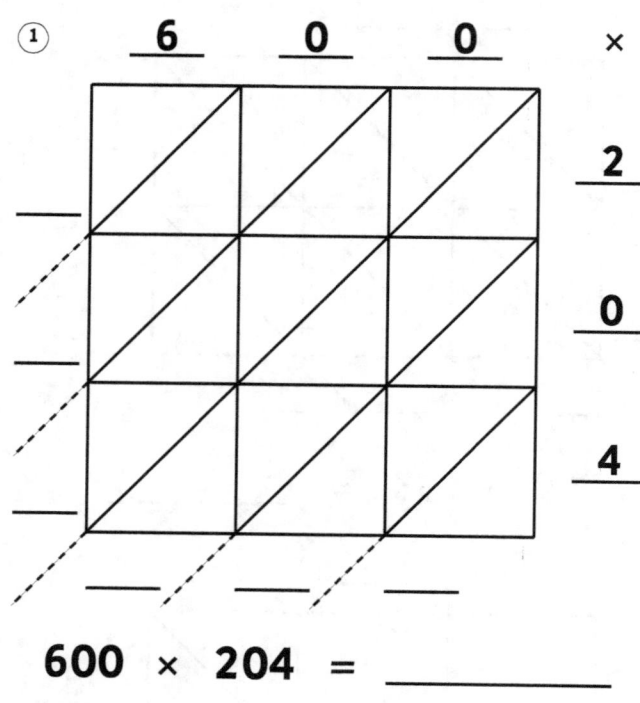

2

0

4

600 × 204 = _____

②   8   7   2   ×

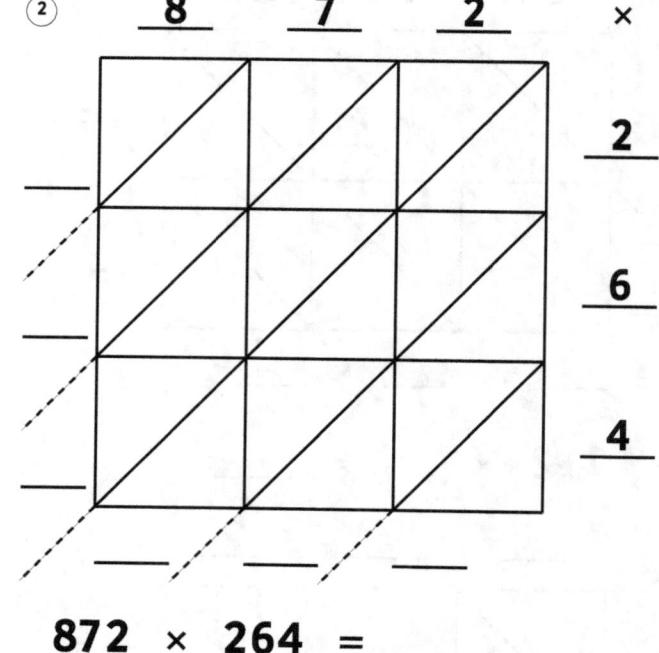

2

6

4

872 × 264 = _____

③   1   0   0   ×

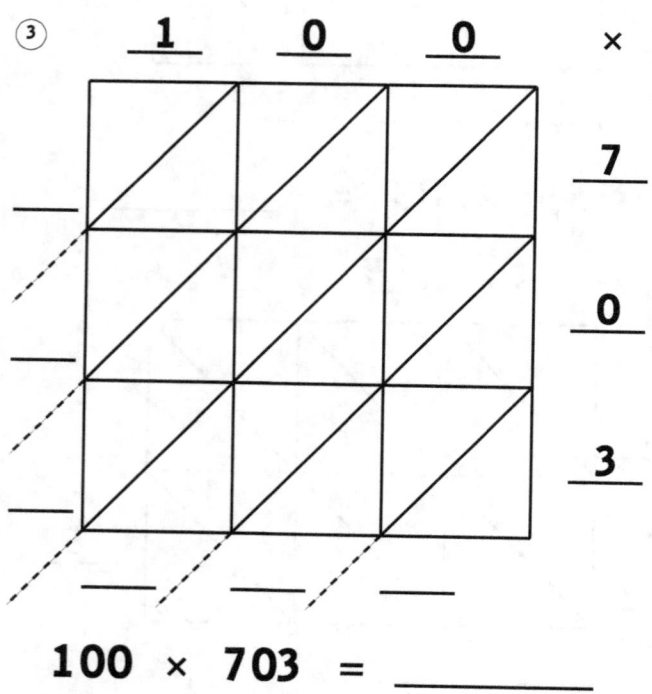

7

0

3

100 × 703 = _____

④   6   1   5   ×

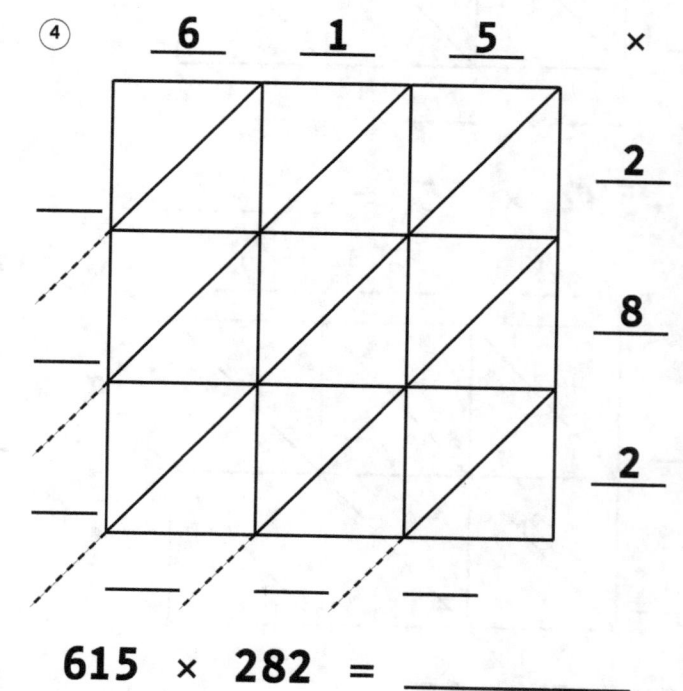

2

8

2

615 × 282 = _____

3 Digits by 3 Digits

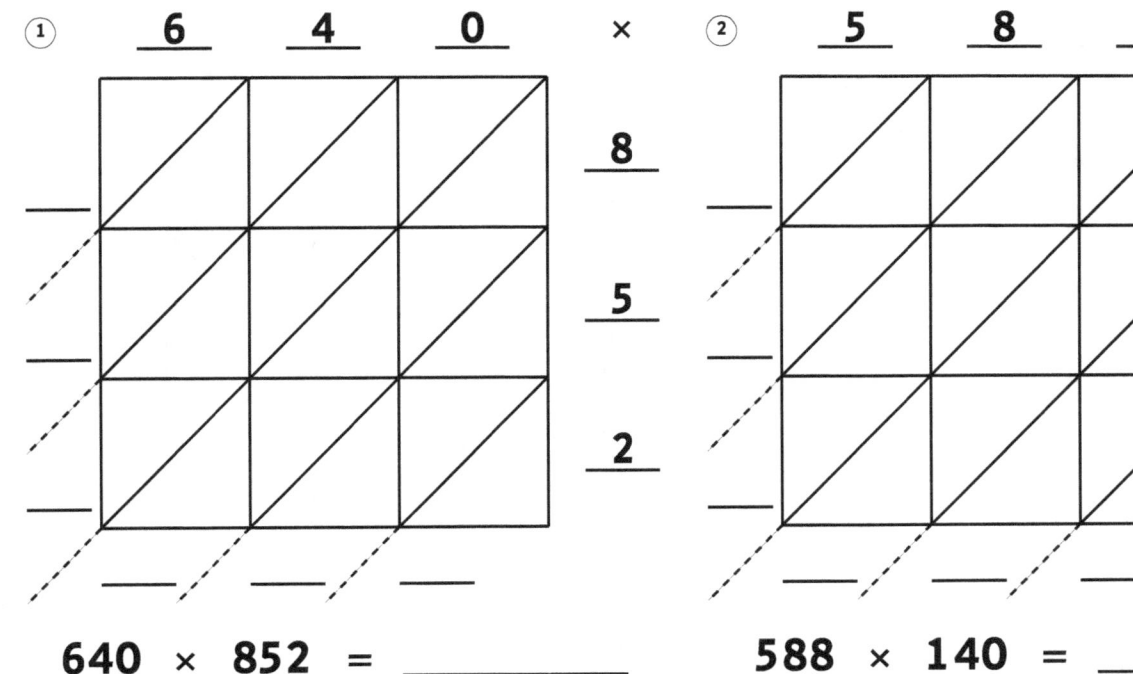

①   6   4   0   ×

8

5

2

640 × 852 = _____

②   5   8   8   ×

1

4

0

588 × 140 = _____

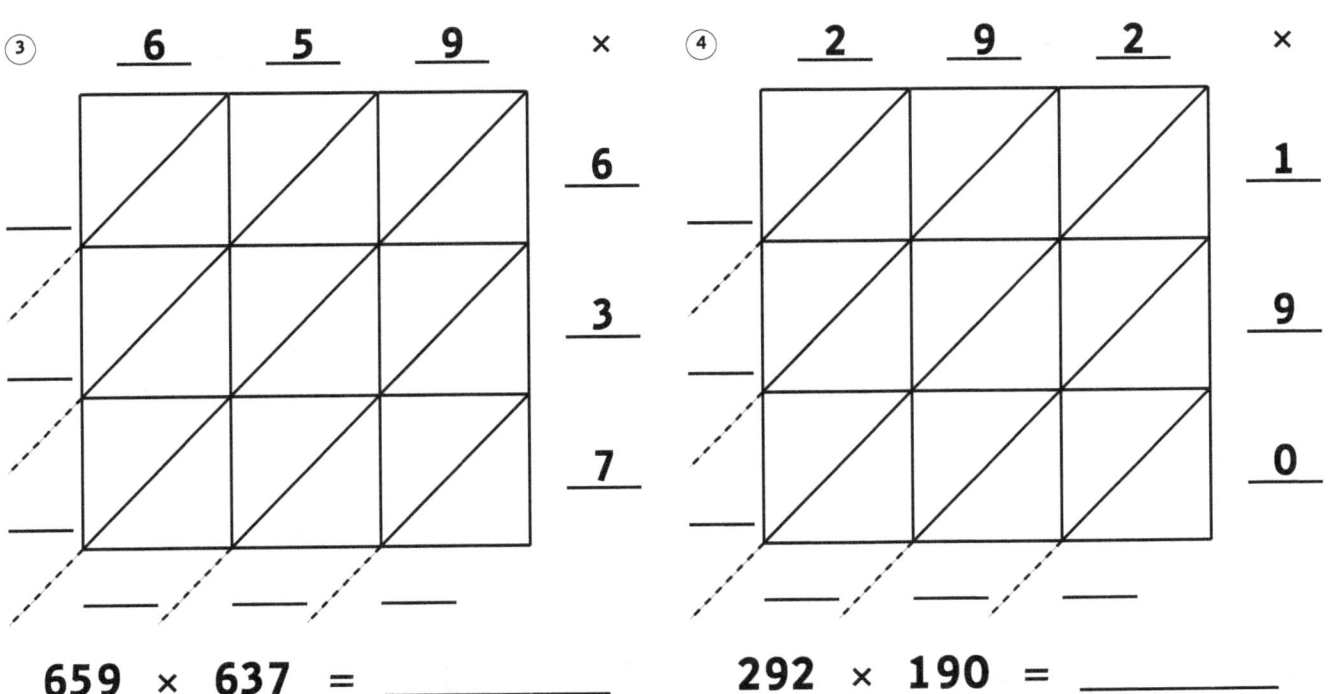

③   6   5   9   ×

6

3

7

659 × 637 = _____

④   2   9   2   ×

1

9

0

292 × 190 = _____

3 Digits by 3 Digits

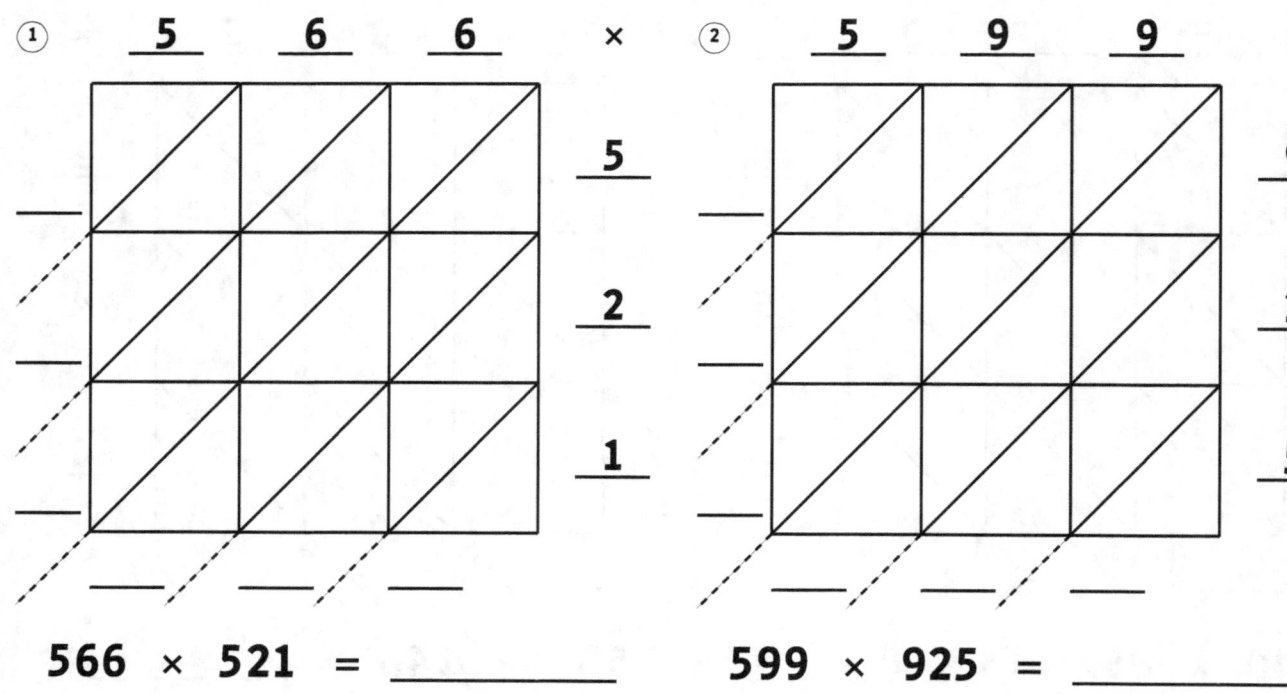

① <u>5</u> <u>6</u> <u>6</u> ×

<u>5</u>

<u>2</u>

<u>1</u>

566 × 521 = _____

② <u>5</u> <u>9</u> <u>9</u> ×

<u>9</u>

<u>2</u>

<u>5</u>

599 × 925 = _____

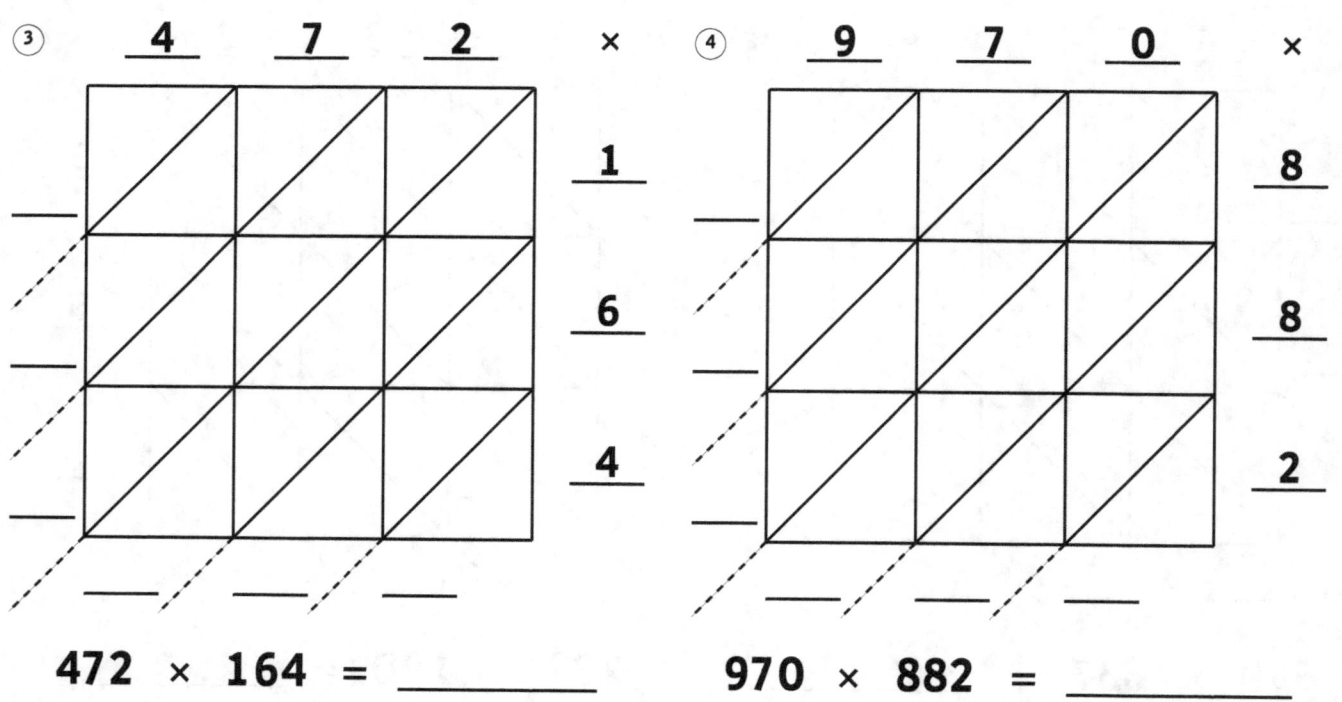

③ <u>4</u> <u>7</u> <u>2</u> ×

<u>1</u>

<u>6</u>

<u>4</u>

472 × 164 = _____

④ <u>9</u> <u>7</u> <u>0</u> ×

<u>8</u>

<u>8</u>

<u>2</u>

970 × 882 = _____

18

3 Digits by 3 Digits

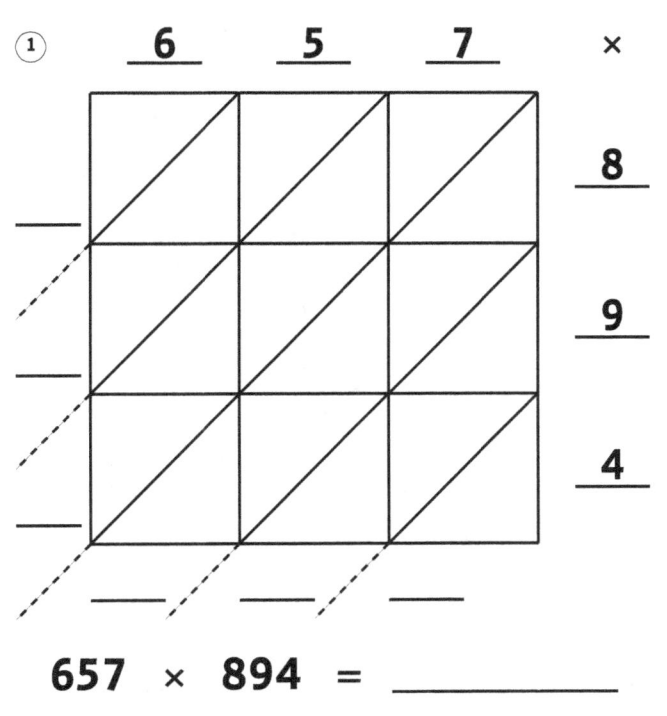

①   6   5   7   ×

8

9

4

657 × 894 = _____

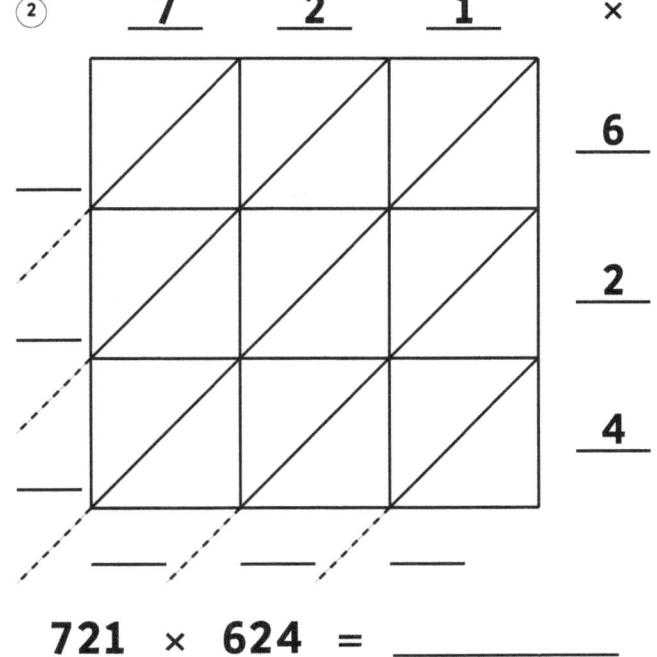

②   7   2   1   ×

6

2

4

721 × 624 = _____

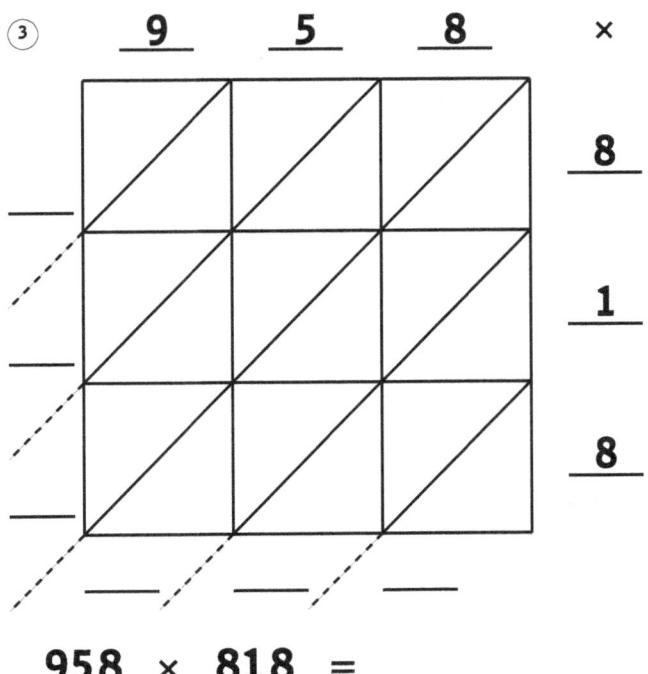

③   9   5   8   ×

8

1

8

958 × 818 = _____

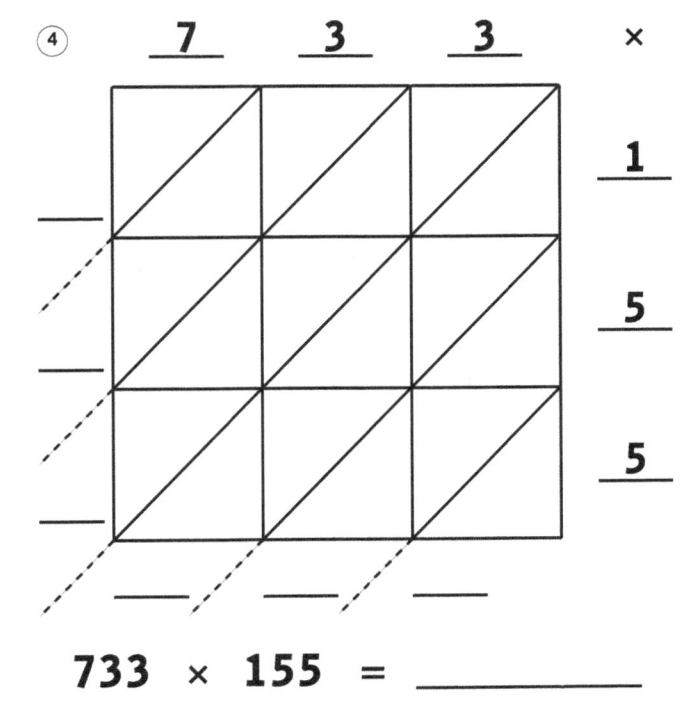

④   7   3   3   ×

1

5

5

733 × 155 = _____

3 Digits by 3 Digits

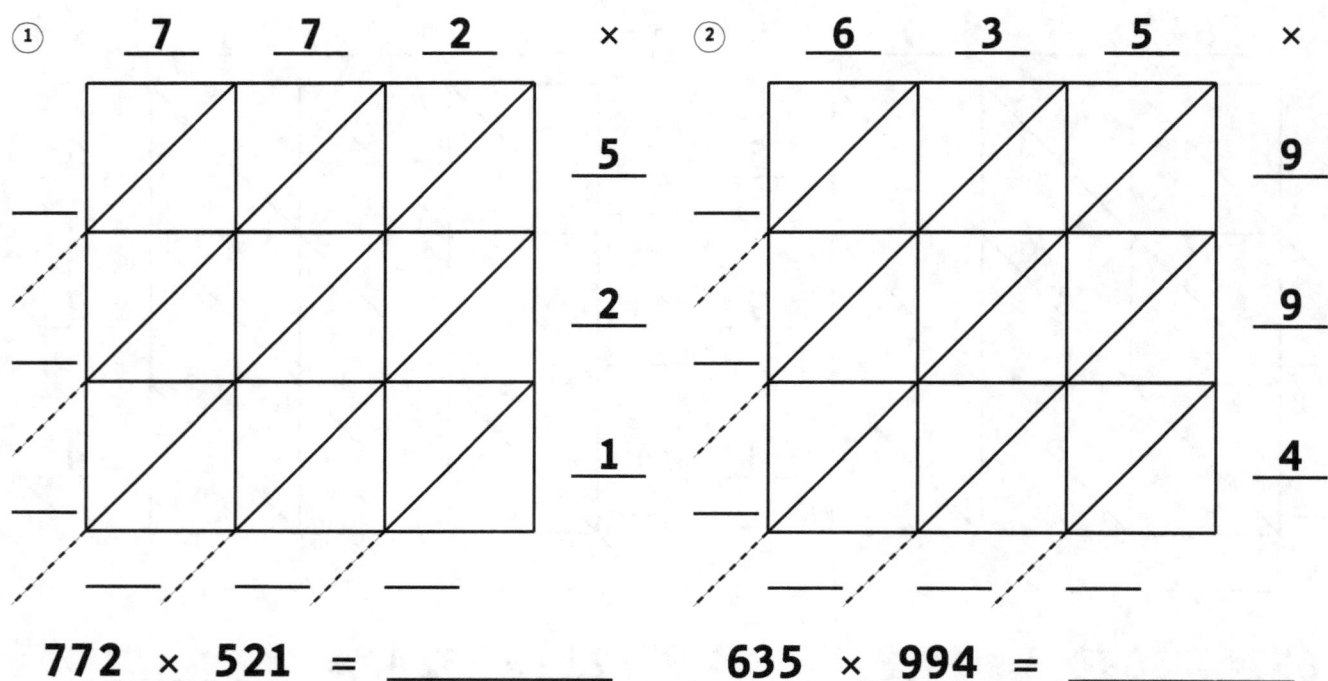

①   7   7   2   ×

5

2

1

772 × 521 = _____

②   6   3   5   ×

9

9

4

635 × 994 = _____

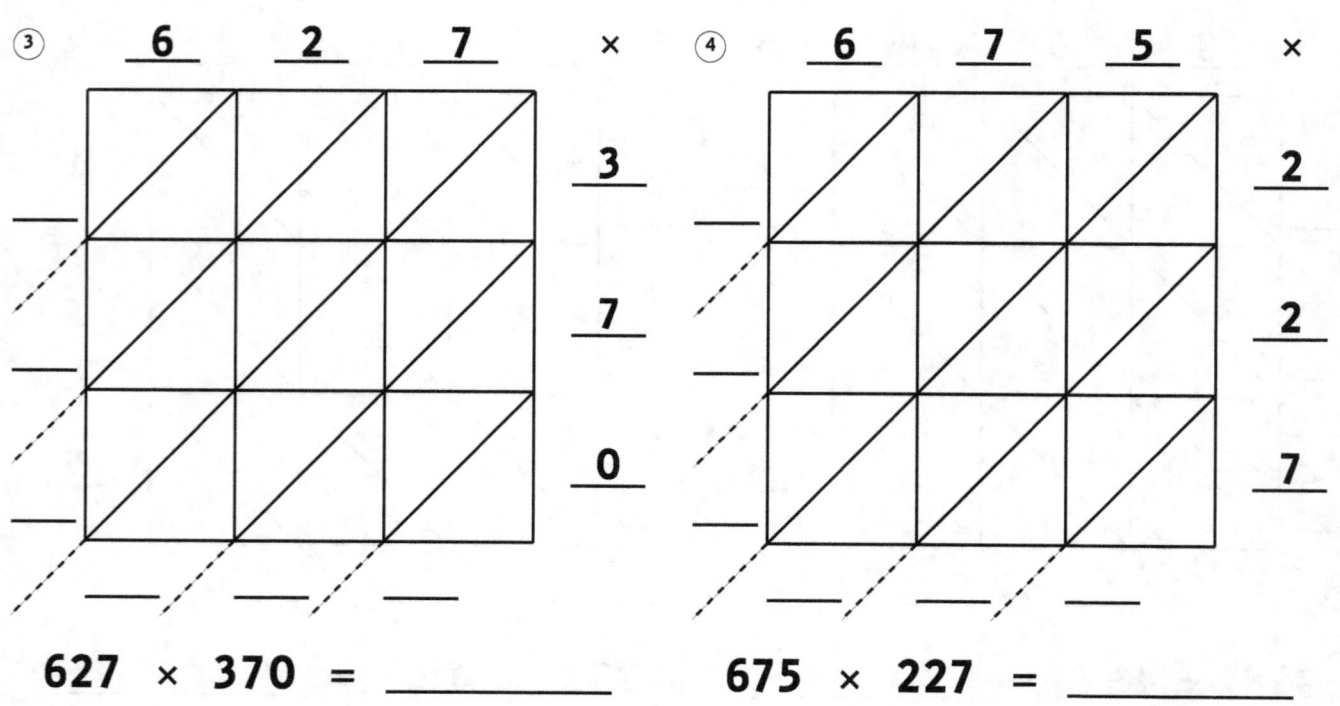

③   6   2   7   ×

3

7

0

627 × 370 = _____

④   6   7   5   ×

2

2

7

675 × 227 = _____

3 Digits by 3 Digits

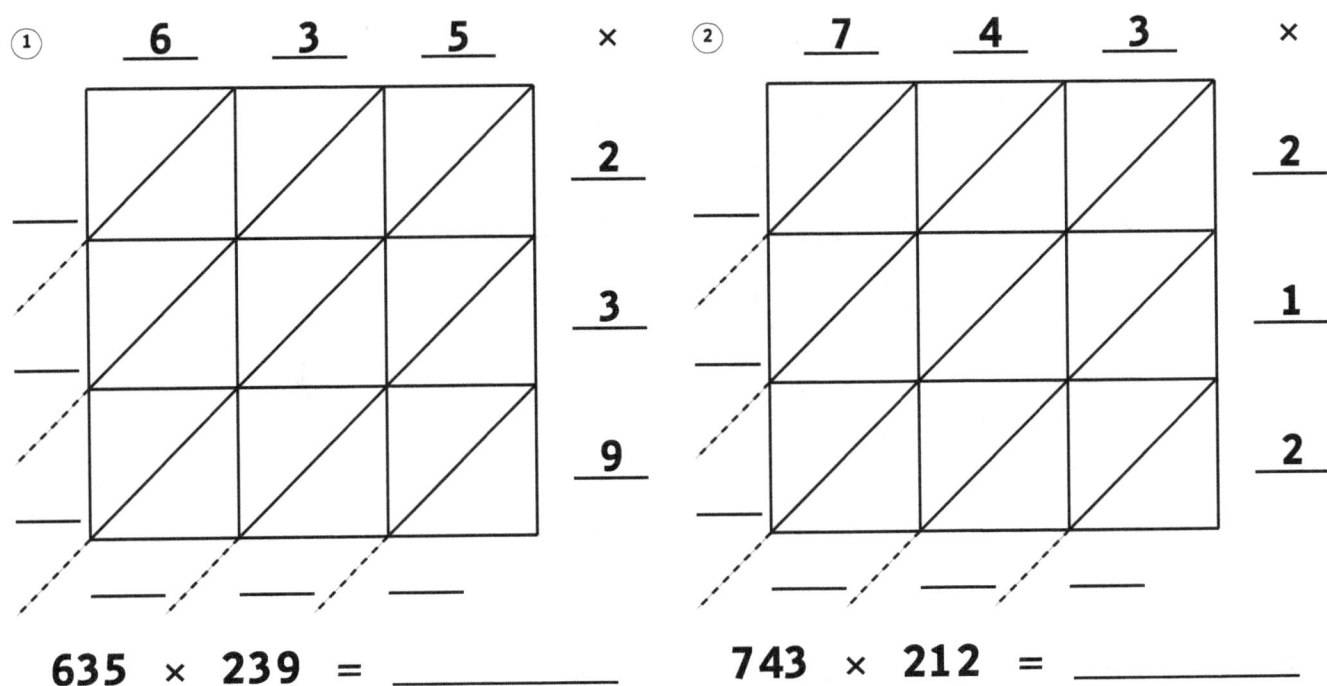

①   6   3   5   ×

2

3

9

635 × 239 = _____

②   7   4   3   ×

2

1

2

743 × 212 = _____

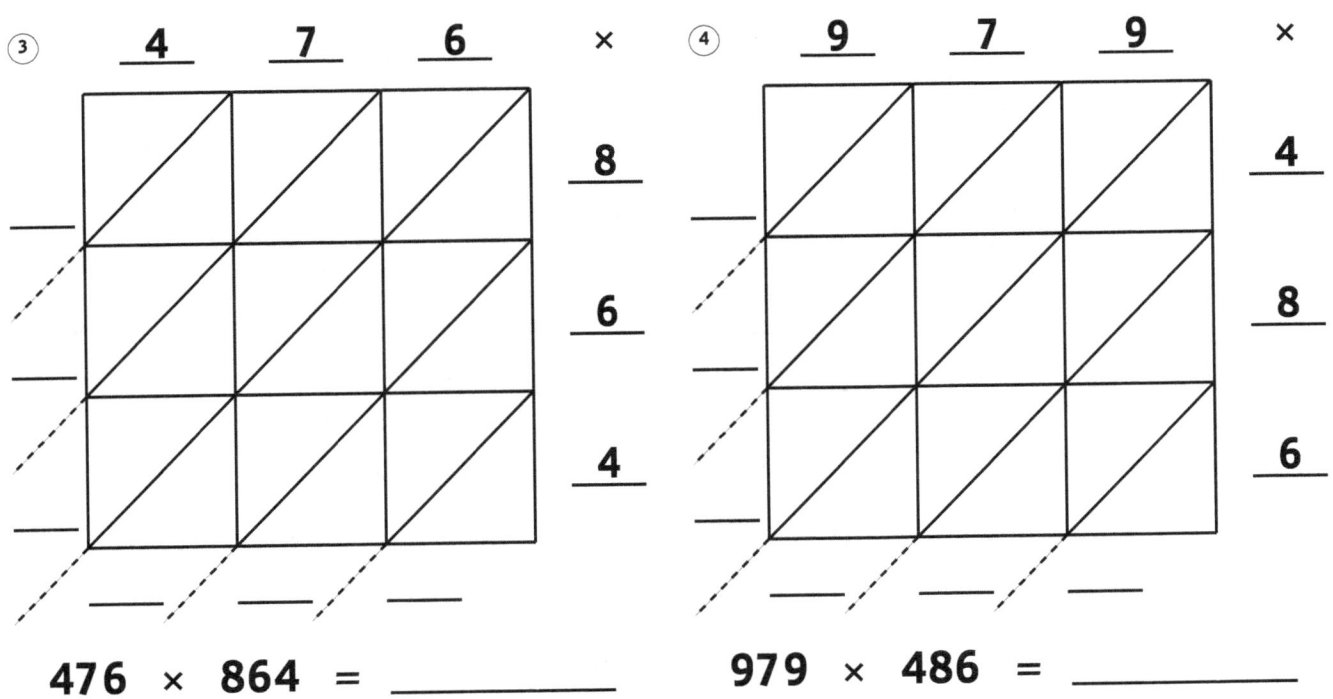

③   4   7   6   ×

8

6

4

476 × 864 = _____

④   9   7   9   ×

4

8

6

979 × 486 = _____

3 Digits by 3 Digits

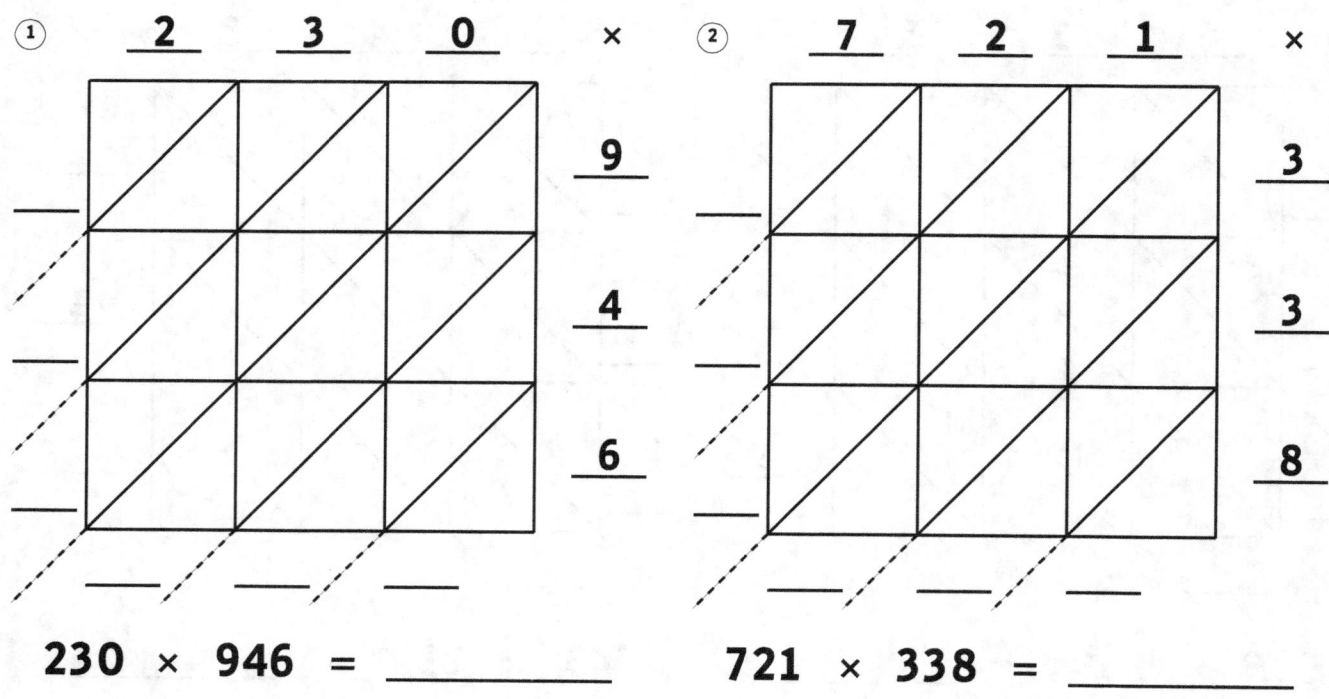

① 2   3   0   ×

9

4

6

230 × 946 = _____

② 7   2   1   ×

3

3

8

721 × 338 = _____

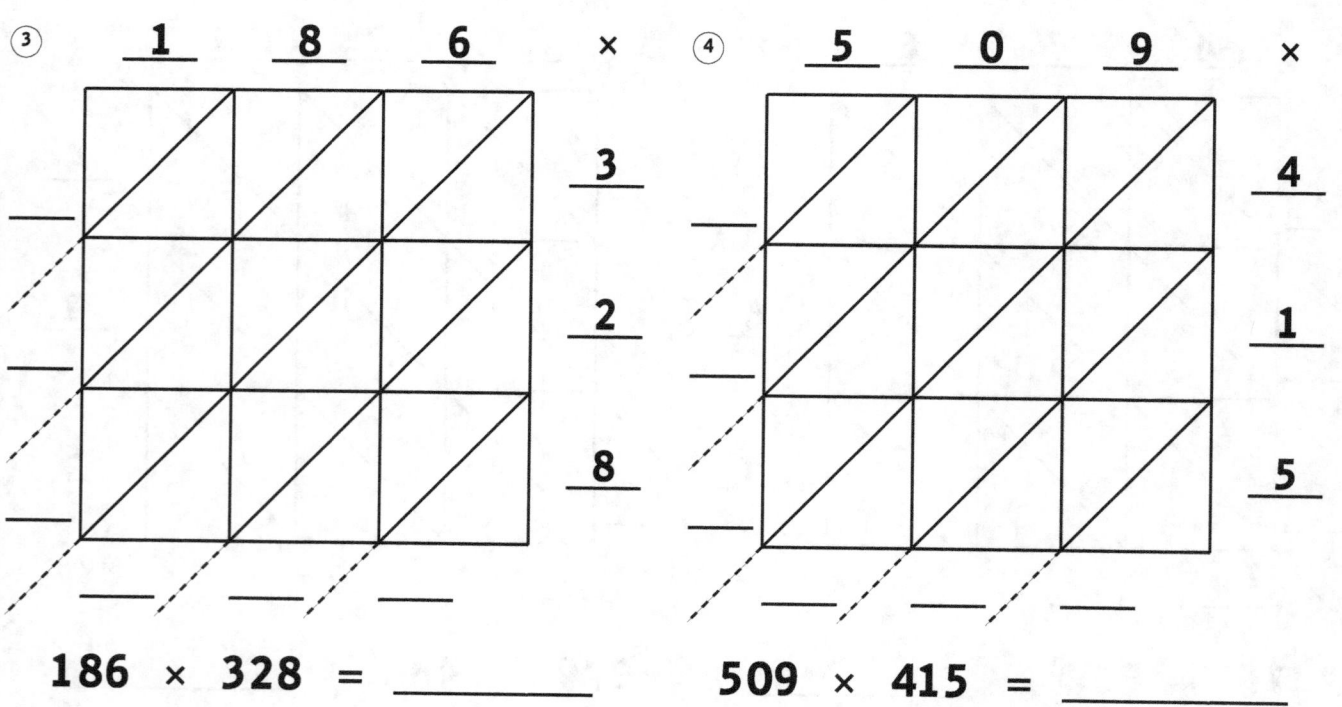

③ 1   8   6   ×

3

2

8

186 × 328 = _____

④ 5   0   9   ×

4

1

5

509 × 415 = _____

22

3 Digits by 3 Digits

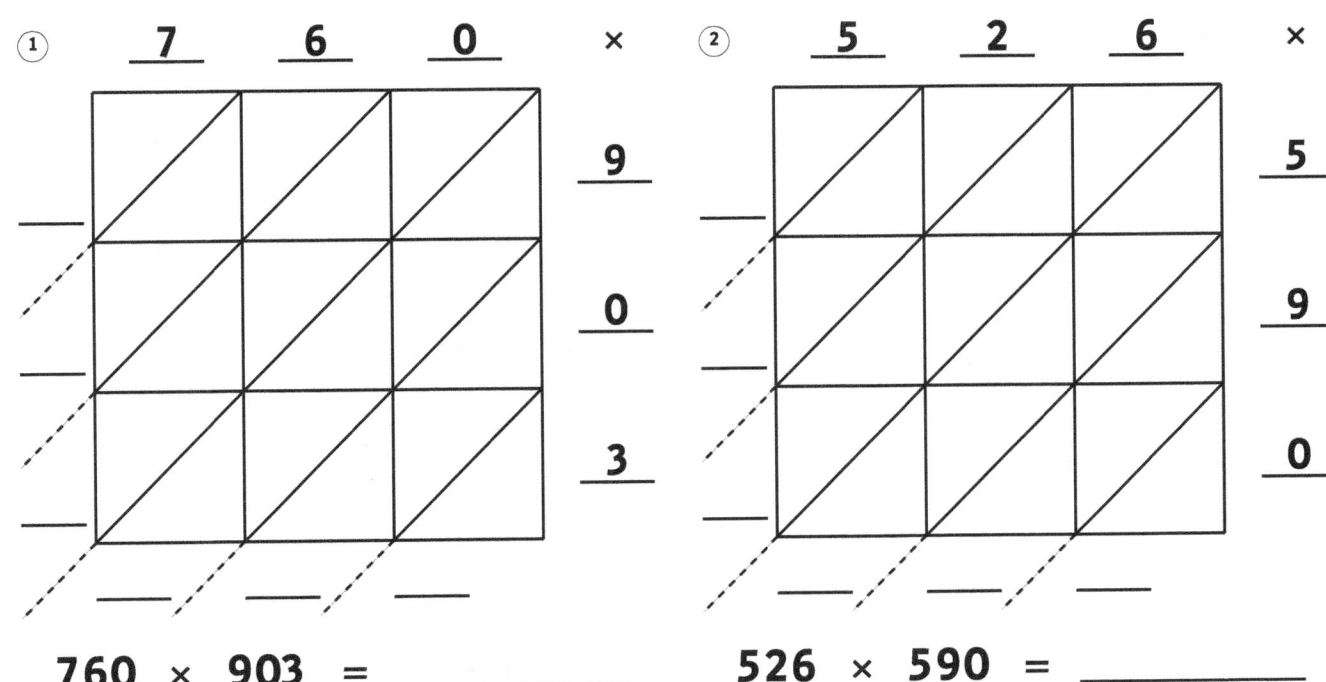

①   7   6   0   ×

9

0

3

760 × 903 = _____

②   5   2   6   ×

5

9

0

526 × 590 = _____

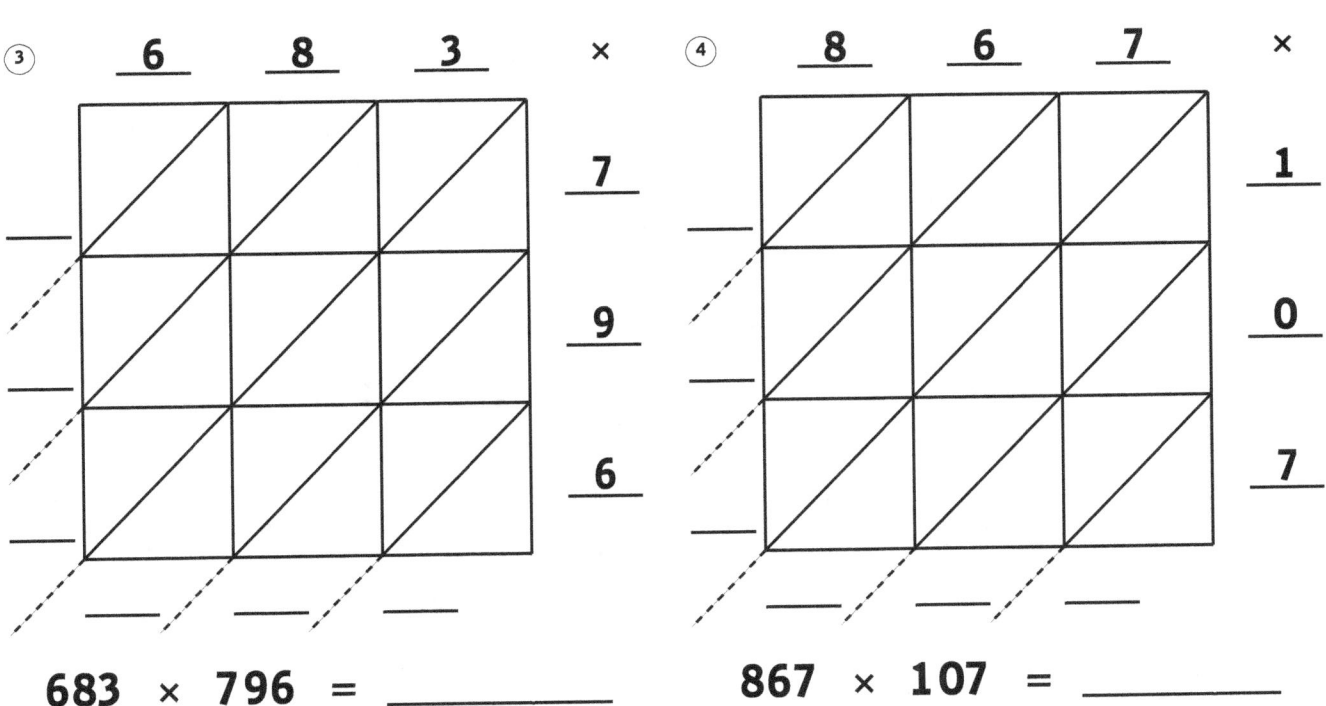

③   6   8   3   ×

7

9

6

683 × 796 = _____

④   8   6   7   ×

1

0

7

867 × 107 = _____

3 Digits by 3 Digits

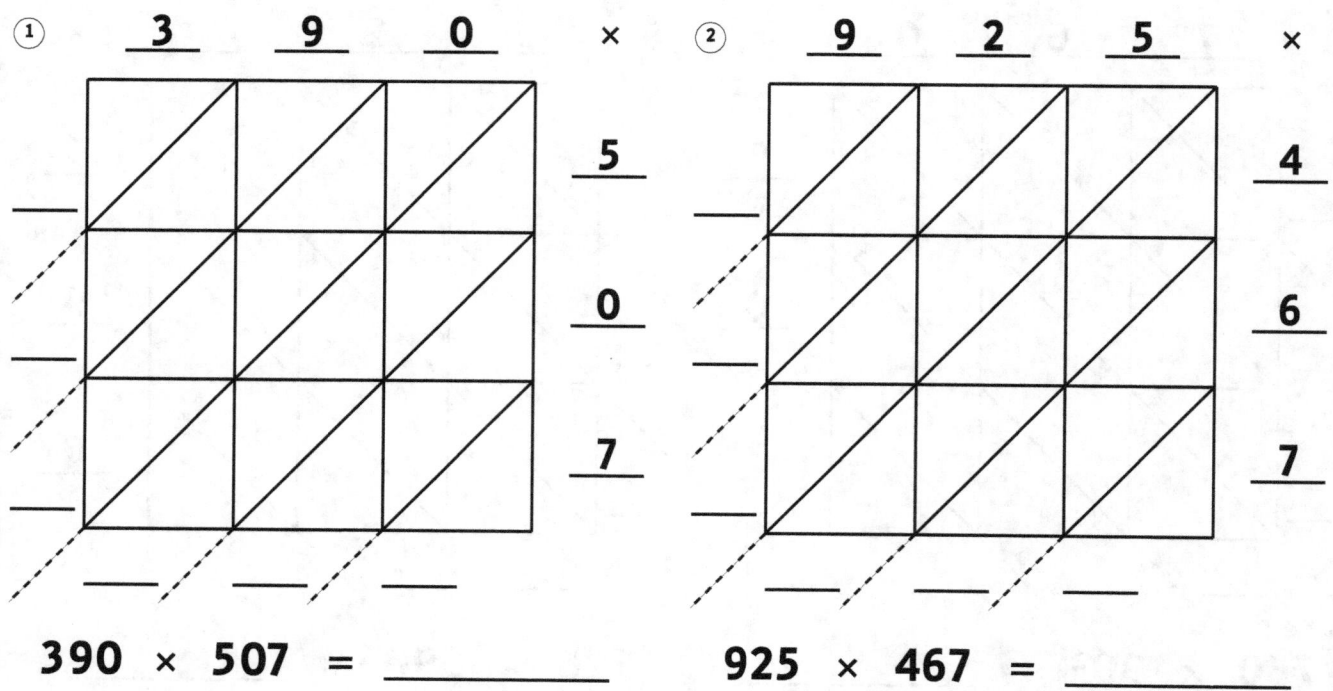

①   3   9   0   ×

5

0

7

390 × 507 = _____

②   9   2   5   ×

4

6

7

925 × 467 = _____

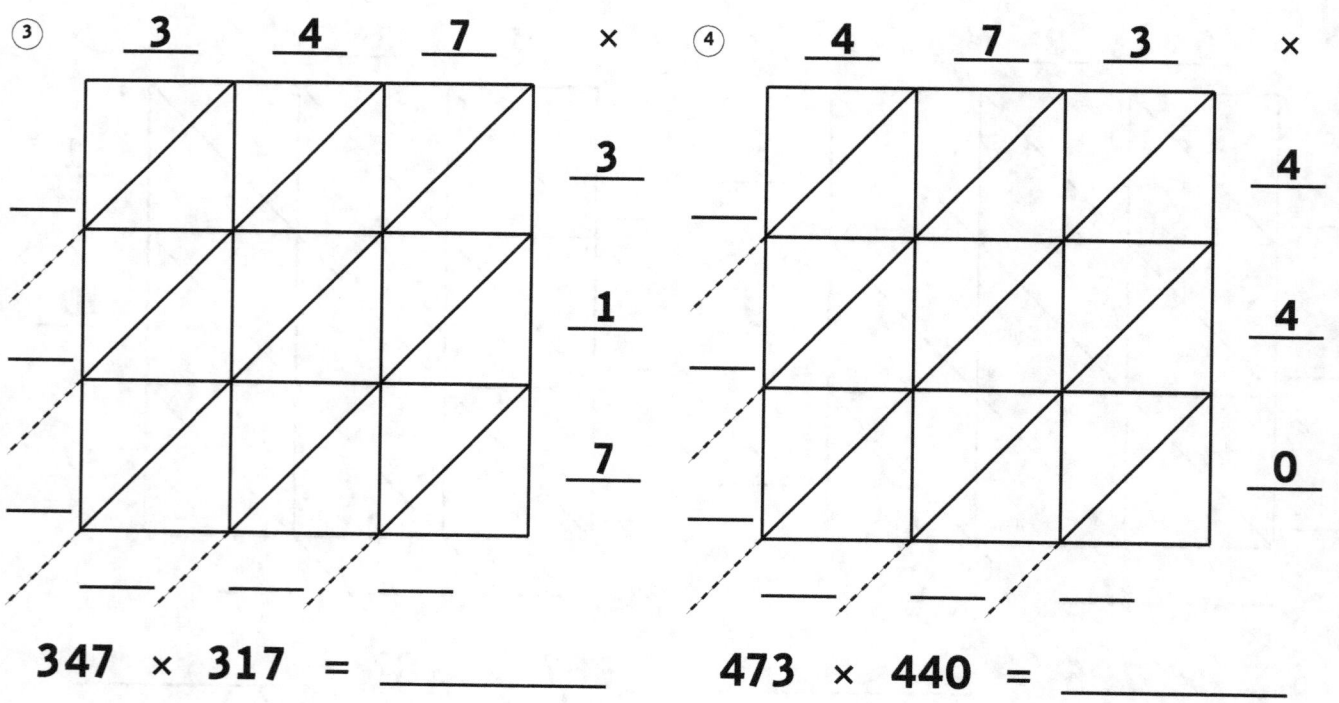

③   3   4   7   ×

3

1

7

347 × 317 = _____

④   4   7   3   ×

4

4

0

473 × 440 = _____

24

3 Digits by 3 Digits

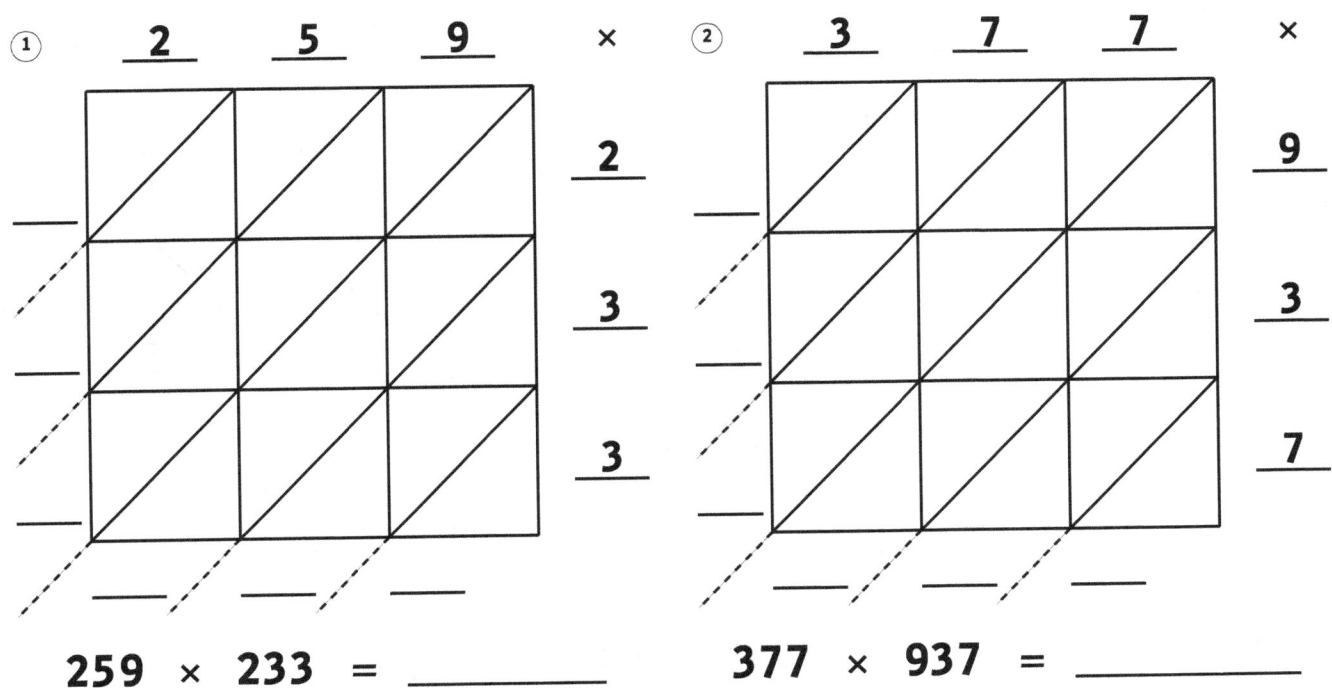

①   2   5   9   ×

               2

               3

               3

259 × 233 = _____

②   3   7   7   ×

               9

               3

               7

377 × 937 = _____

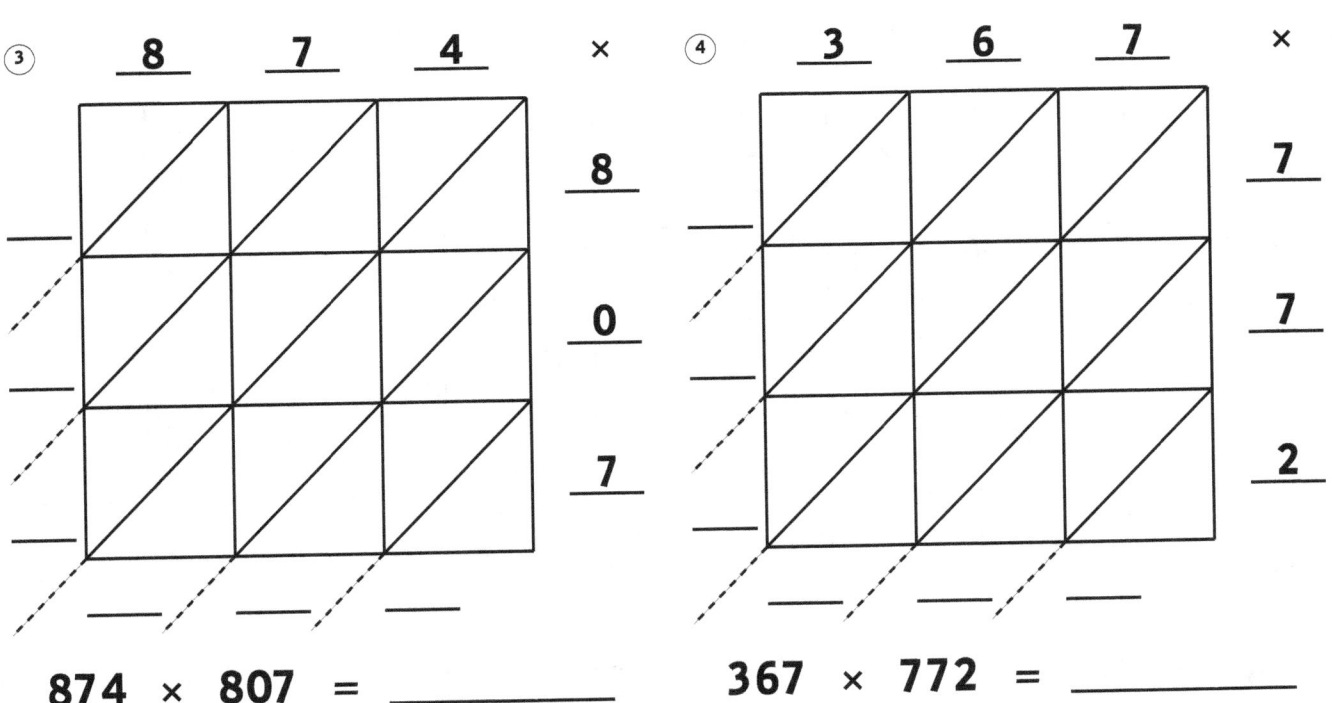

③   8   7   4   ×

               8

               0

               7

874 × 807 = _____

④   3   6   7   ×

               7

               7

               2

367 × 772 = _____

25

3 Digits by 3 Digits

①   4   6   8   ×

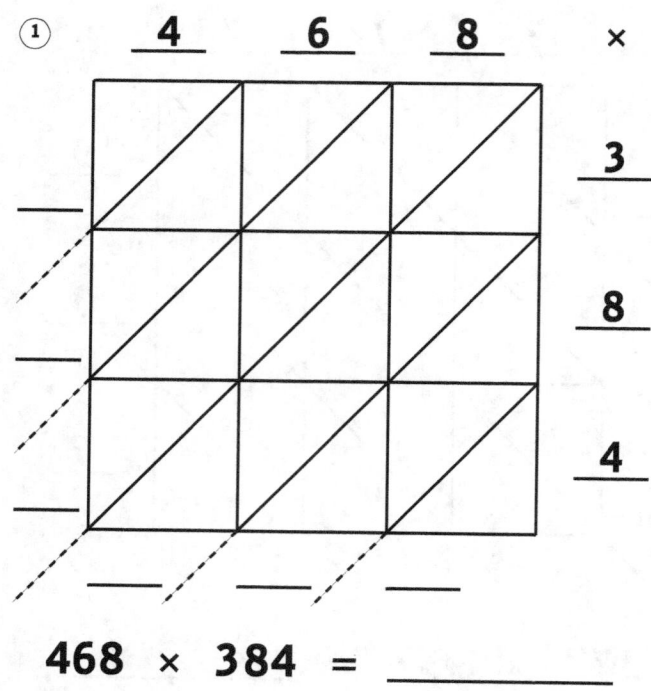

3

8

4

468 × 384 = _____

②   6   8   6   ×

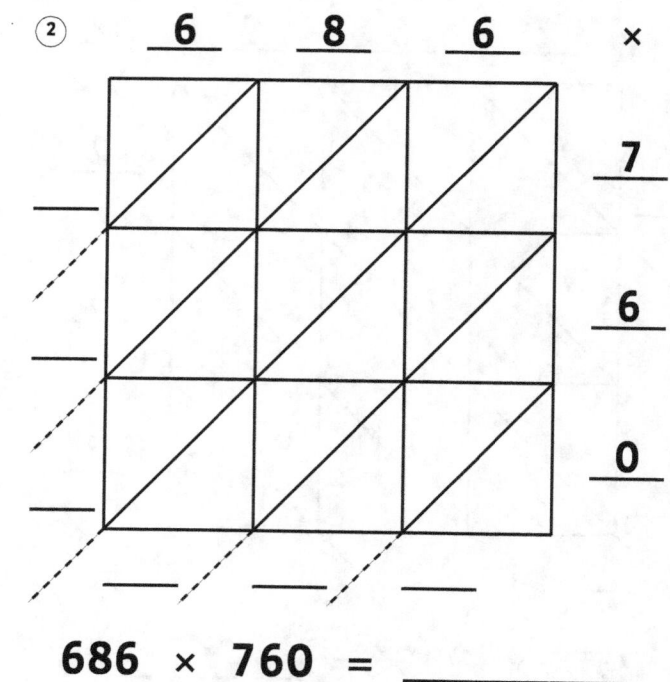

7

6

0

686 × 760 = _____

③   2   0   6   ×

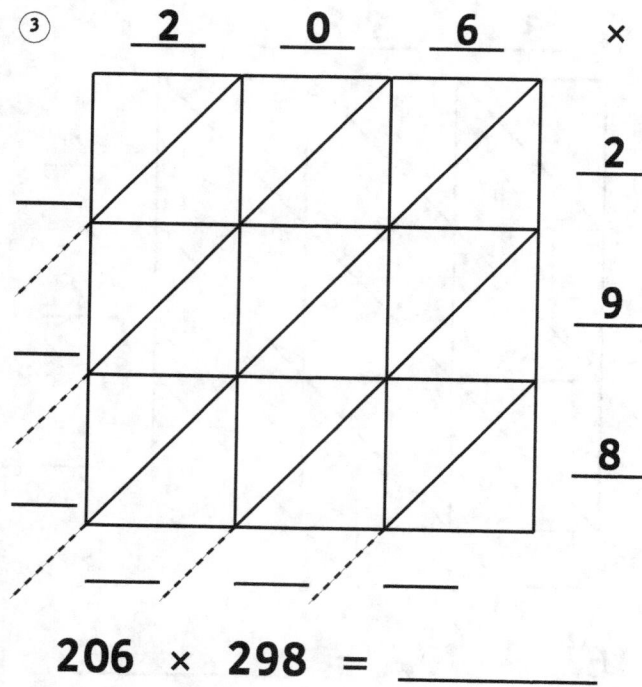

2

9

8

206 × 298 = _____

④   2   7   5   ×

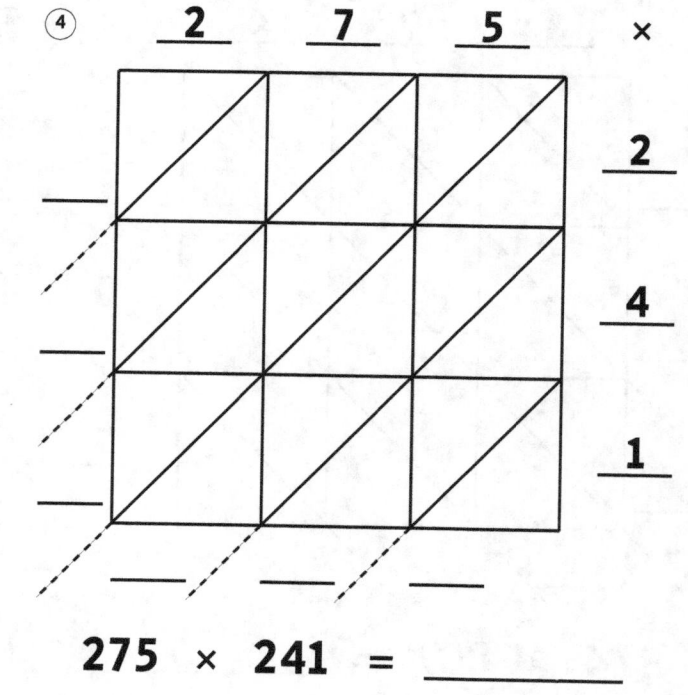

2

4

1

275 × 241 = _____

3 Digits by 3 Digits

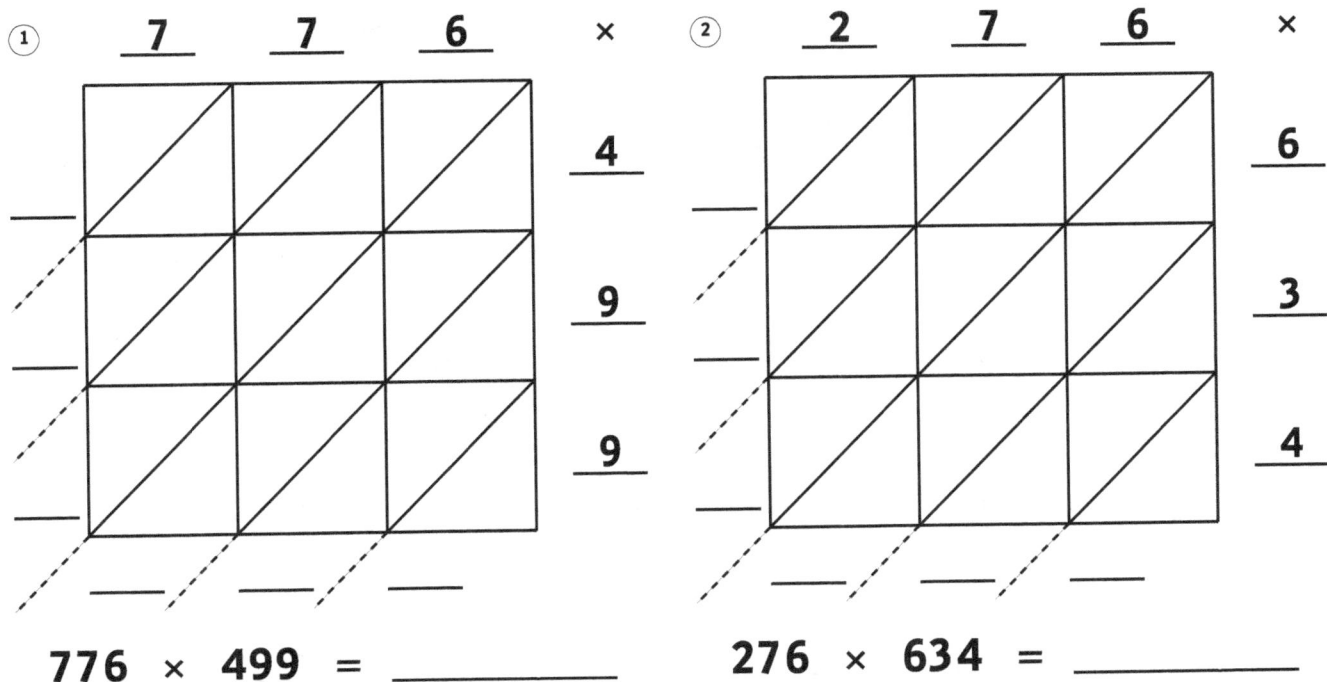

①   7   7   6   ×

4

9

9

776 × 499 = _____

②   2   7   6   ×

6

3

4

276 × 634 = _____

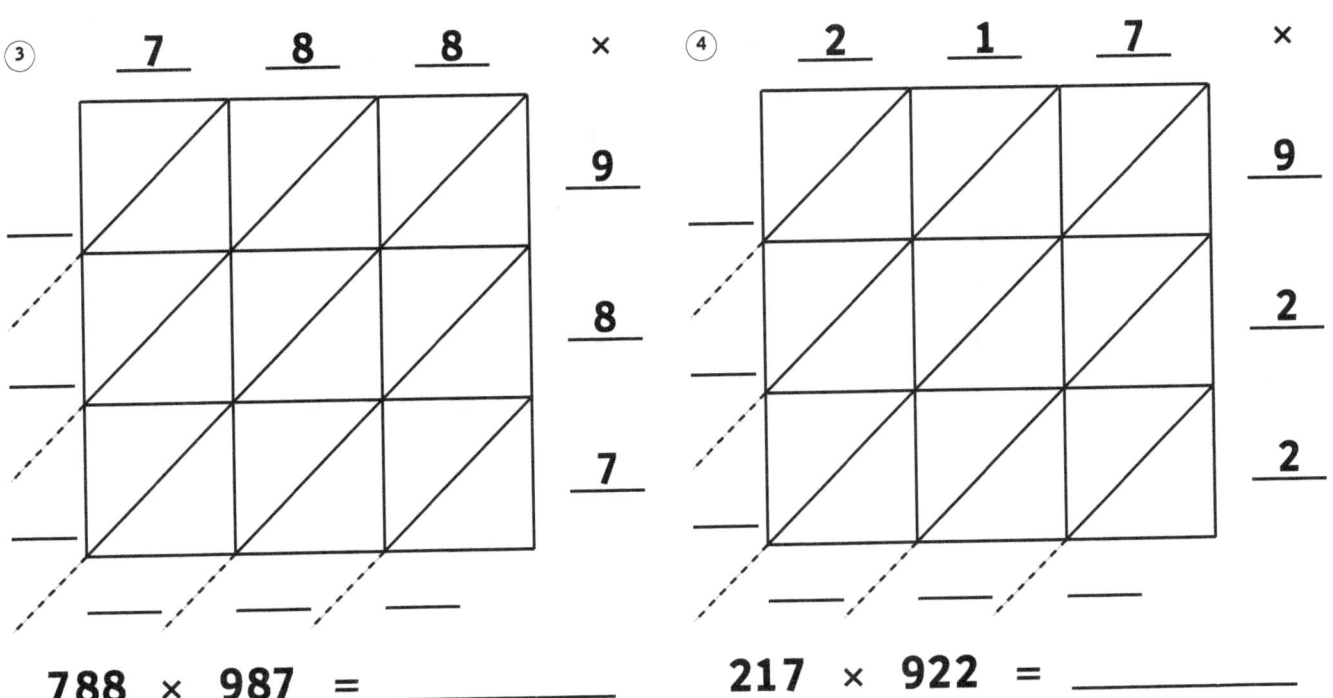

③   7   8   8   ×

9

8

7

788 × 987 = _____

④   2   1   7   ×

9

2

2

217 × 922 = _____

3 Digits by 3 Digits

①   9   1   0   ×

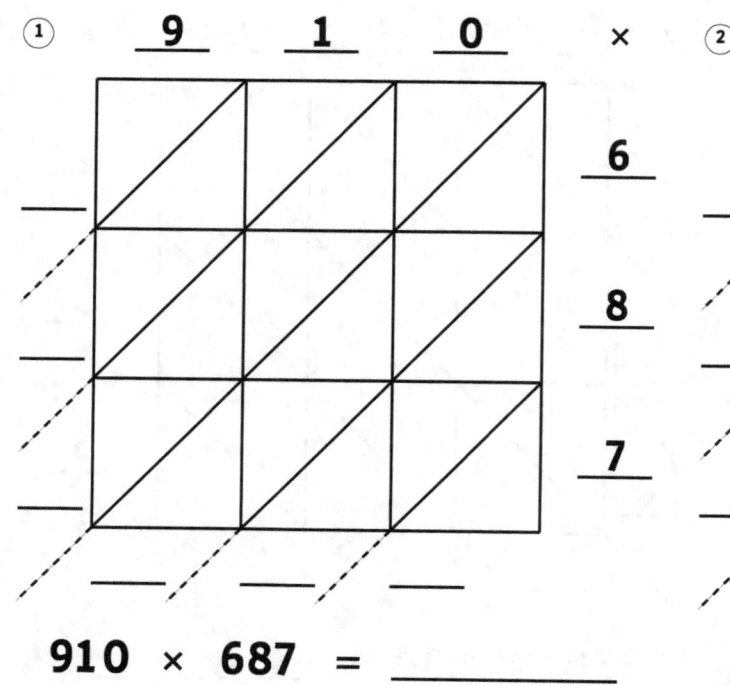

910 × 687 = _____

②   5   6   2   ×

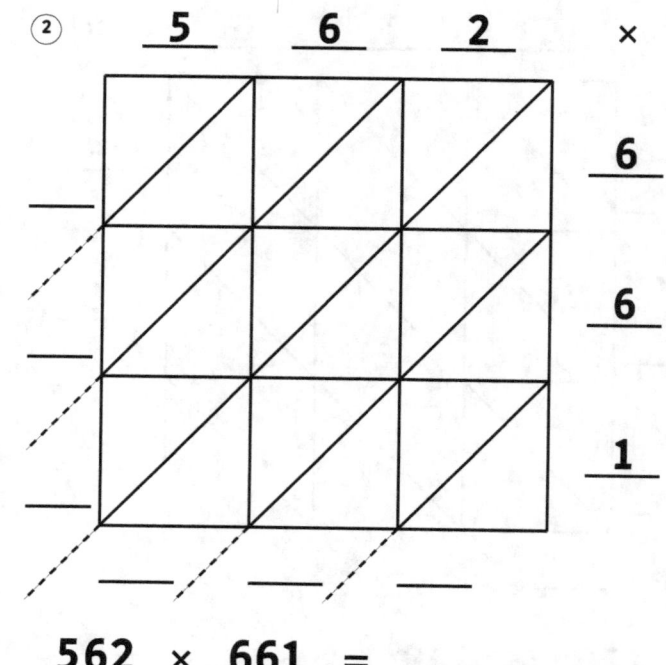

562 × 661 = _____

③   6   2   3   ×

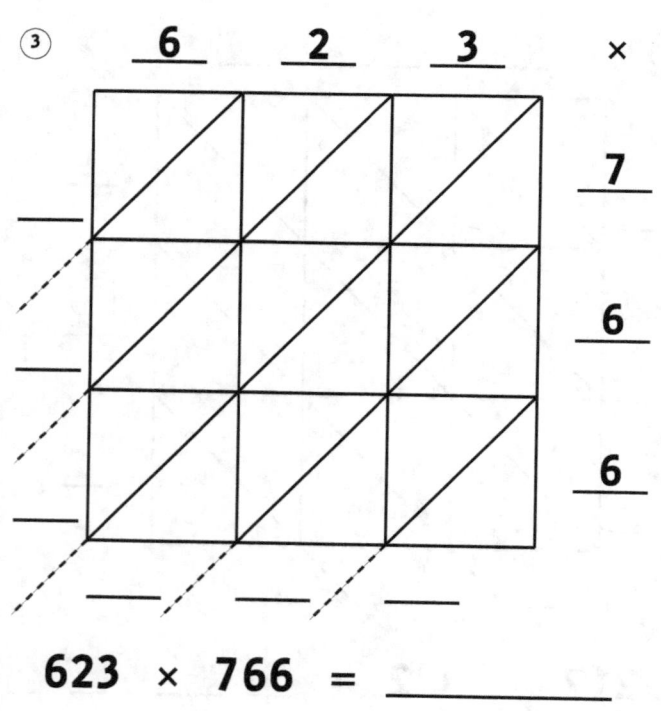

623 × 766 = _____

④   8   8   6   ×

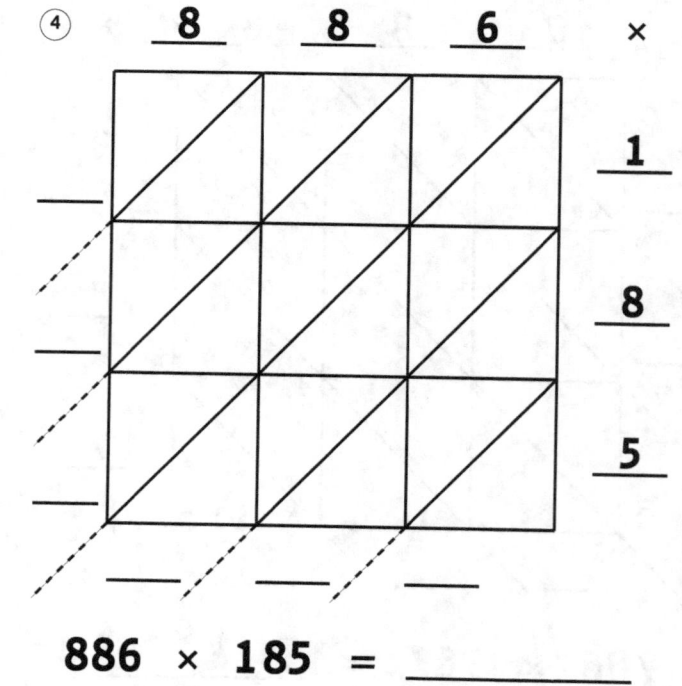

886 × 185 = _____

3 Digits by 3 Digits

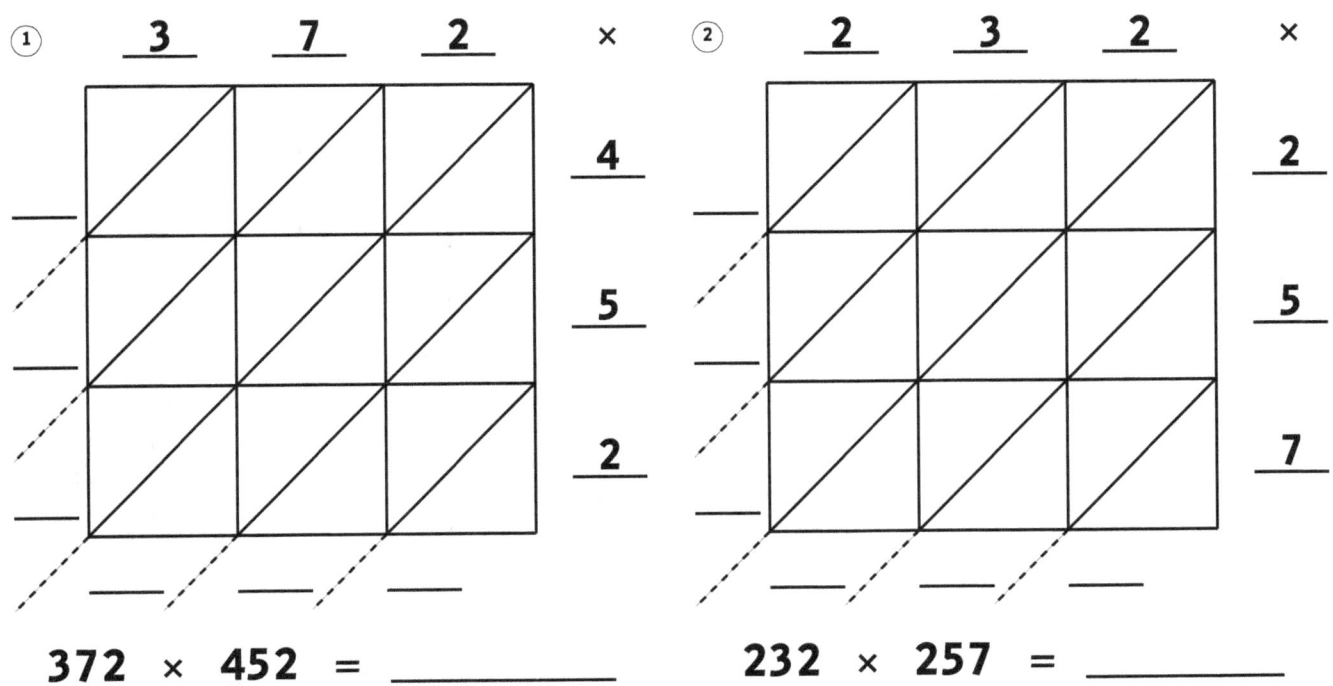

① 3   7   2   ×
4
5
2

372 × 452 = _____

② 2   3   2   ×
2
5
7

232 × 257 = _____

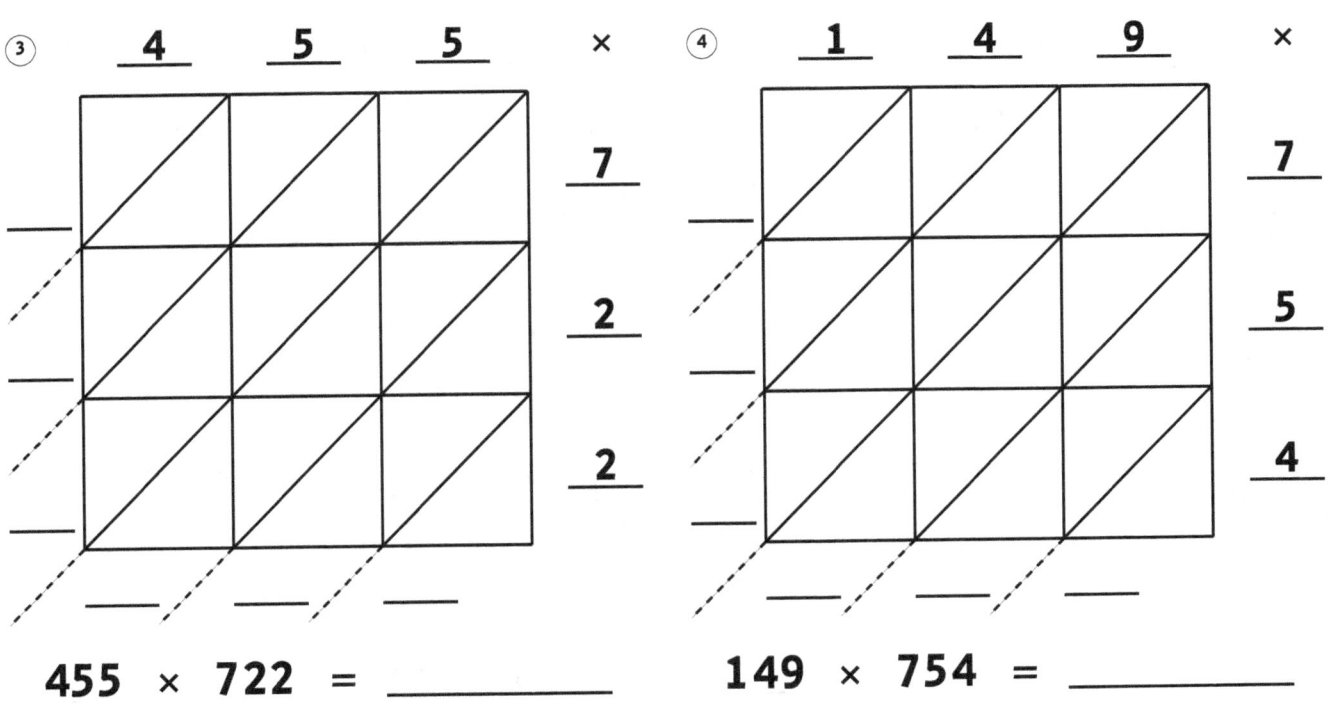

③ 4   5   5   ×
7
2
2

455 × 722 = _____

④ 1   4   9   ×
7
5
4

149 × 754 = _____

3 Digits by 3 Digits

①   9   4   8   ×

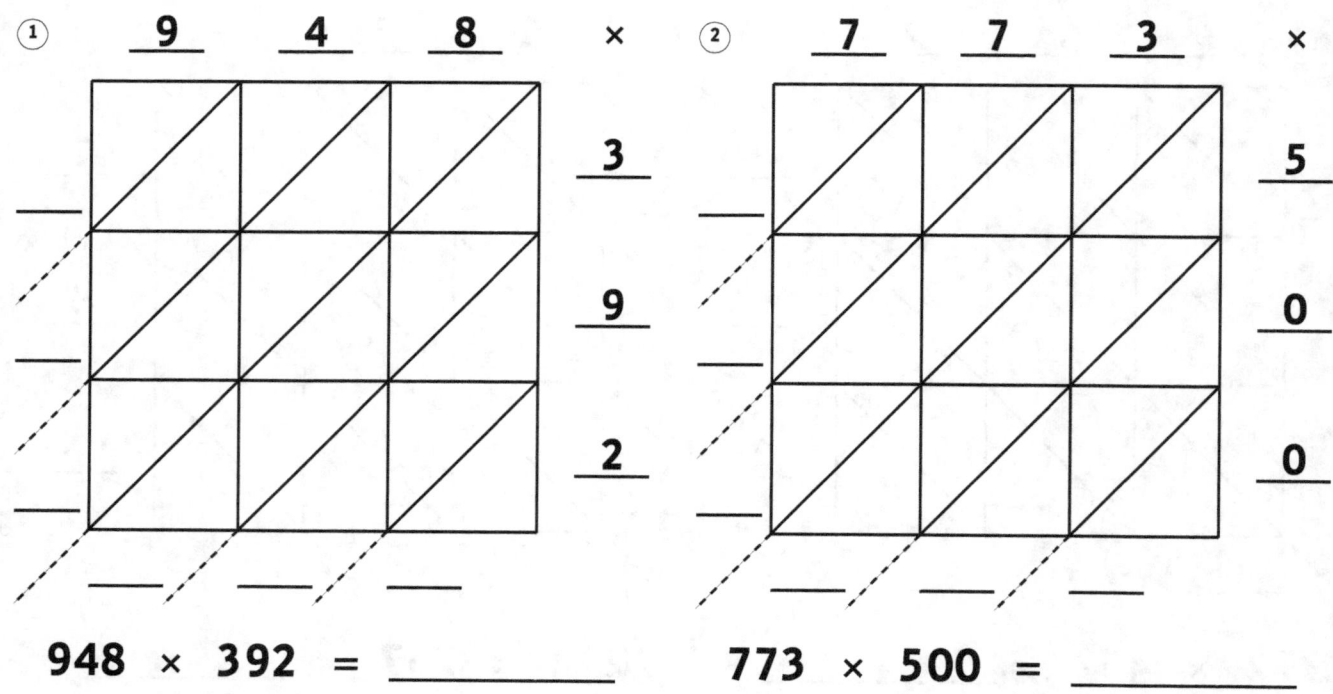

3

9

2

948 × 392 = _____

②   7   7   3   ×

5

0

0

773 × 500 = _____

③   8   4   2   ×

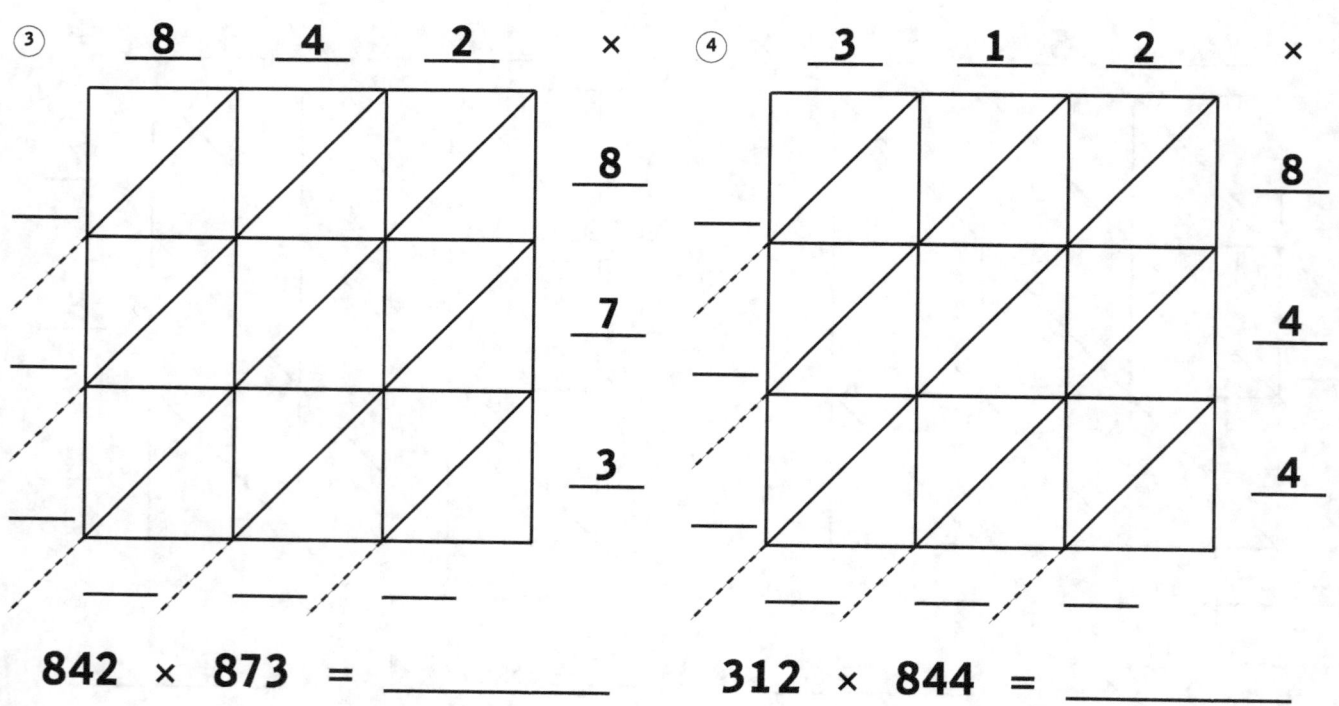

8

7

3

842 × 873 = _____

④   3   1   2   ×

8

4

4

312 × 844 = _____

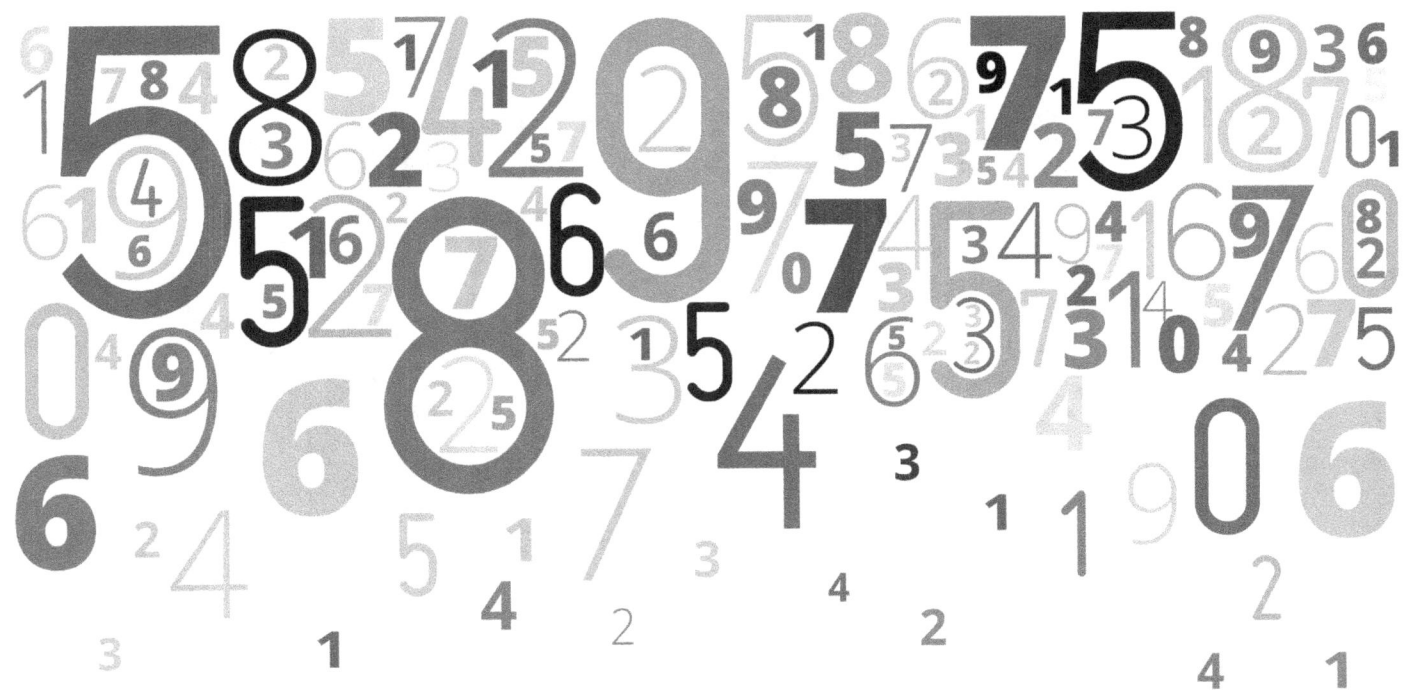

Section Two
Division

3 Digits by 1 Digit

① 8 ⟌2 4 3 ⟍R

② 3 ⟌8 1 8 ⟍R

③ 3 ⟌3 2 2 ⟍R

④ 3 ⟌8 5 9 ⟍R

⑤ 3 ⟌1 8 5 ⟍R

⑥ 5 ⟌1 5 6 ⟍R

⑦ 6 ⟌7 4 7 ⟍R

⑧ 5 ⟌5 1 0 ⟍R

⑨ 2 ⟌7 2 0 ⟍R

32

3 Digits by 1 Digit

① $5 \overline{)8 \quad 4 \quad 4}$ R

② $4 \overline{)3 \quad 9 \quad 8}$ R

③ $7 \overline{)3 \quad 1 \quad 3}$ R

④ $5 \overline{)8 \quad 9 \quad 4}$ R

⑤ $2 \overline{)4 \quad 7 \quad 3}$ R

⑥ $7 \overline{)8 \quad 8 \quad 7}$ R

⑦ $9 \overline{)6 \quad 6 \quad 4}$ R

⑧ $7 \overline{)7 \quad 8 \quad 7}$ R

⑨ $4 \overline{)2 \quad 2 \quad 3}$ R

3 Digits by 1 Digit

①
$$6 \overline{)716} \,^R$$

②
$$5 \overline{)477} \,^R$$

③
$$2 \overline{)875} \,^R$$

④
$$9 \overline{)262} \,^R$$

⑤
$$9 \overline{)873} \,^R$$

⑥
$$3 \overline{)942} \,^R$$

⑦
$$6 \overline{)795} \,^R$$

⑧
$$8 \overline{)128} \,^R$$

⑨
$$9 \overline{)693} \,^R$$

3 Digits by 1 Digit

①
$$4 \overline{) \, 9 \quad 4 \quad 8} \, ^R$$

②
$$5 \overline{) \, 5 \quad 4 \quad 9} \, ^R$$

③
$$7 \overline{) \, 5 \quad 6 \quad 2} \, ^R$$

④
$$9 \overline{) \, 6 \quad 9 \quad 4} \, ^R$$

⑤
$$9 \overline{) \, 3 \quad 4 \quad 9} \, ^R$$

⑥
$$2 \overline{) \, 7 \quad 4 \quad 4} \, ^R$$

⑦
$$2 \overline{) \, 6 \quad 6 \quad 8} \, ^R$$

⑧
$$2 \overline{) \, 7 \quad 8 \quad 7} \, ^R$$

⑨
$$3 \overline{) \, 5 \quad 9 \quad 2} \, ^R$$

3 Digits by 1 Digit

① 4) 8 2 3 ⎤R

② 8) 4 6 9 ⎤R

③ 6) 3 9 9 ⎤R

④ 3) 5 8 2 ⎤R

⑤ 5) 9 8 0 ⎤R

⑥ 7) 8 6 9 ⎤R

⑦ 6) 1 5 1 ⎤R

⑧ 8) 8 3 5 ⎤R

⑨ 4) 2 1 6 ⎤R

36

3 Digits by 1 Digit

①

3⟌9 1 4 ᴿ

②

4⟌7 9 5 ᴿ

③

2⟌9 0 8 ᴿ

④

3⟌5 1 0 ᴿ

⑤

2⟌7 5 0 ᴿ

⑥

7⟌4 3 4 ᴿ

⑦

9⟌8 3 6 ᴿ

⑧

7⟌3 4 5 ᴿ

⑨

6⟌1 5 9 ᴿ

3 Digits by 1 Digit

①
$$3 \overline{)334} ^R$$

②
$$2 \overline{)693} ^R$$

③
$$8 \overline{)916} ^R$$

④
$$7 \overline{)567} ^R$$

⑤
$$6 \overline{)519} ^R$$

⑥
$$2 \overline{)821} ^R$$

⑦
$$3 \overline{)530} ^R$$

⑧
$$9 \overline{)407} ^R$$

⑨
$$5 \overline{)196} ^R$$

38

3 Digits by 1 Digit

(1)

$$3 \overline{)504} \text{ R}$$

(2)

$$3 \overline{)521} \text{ R}$$

(3)

$$9 \overline{)874} \text{ R}$$

(4)

$$2 \overline{)283} \text{ R}$$

(5)

$$9 \overline{)767} \text{ R}$$

(6)

$$5 \overline{)438} \text{ R}$$

(7)

$$6 \overline{)434} \text{ R}$$

(8)

$$2 \overline{)444} \text{ R}$$

(9)

$$2 \overline{)916} \text{ R}$$

3 Digits by 2 Digits

① 1 7) 9 0 4   R

② 7 6) 1 9 4   I

③ 8 7) 2 9 6   R

④ 8 3) 9 8 3   I

3 Digits by 2 Digits

①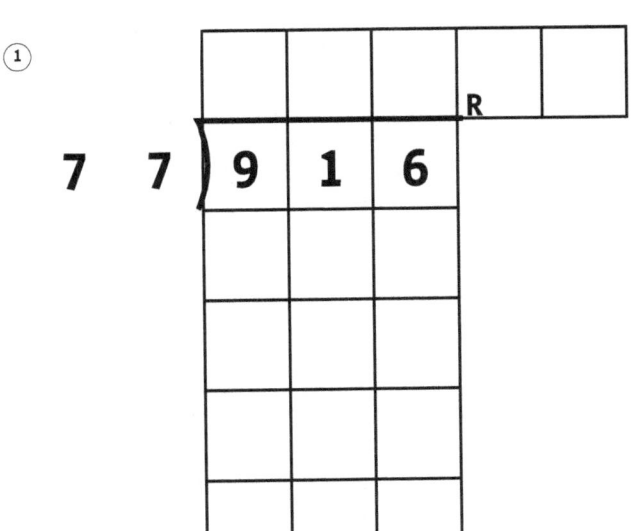

$$7\ 7\ )\ \overline{9\ 1\ 6}\quad R$$

②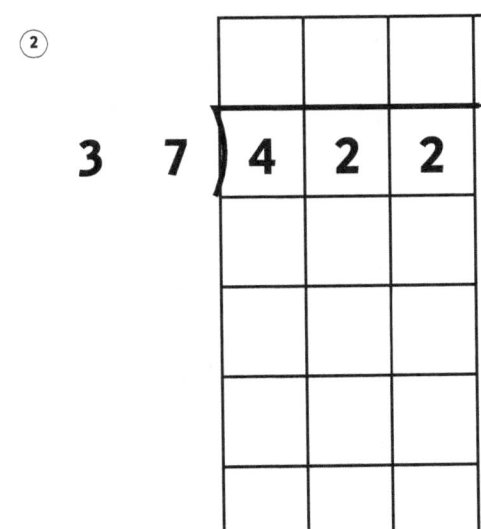

$$3\ 7\ )\ \overline{4\ 2\ 2}$$

③

$$8\ 2\ )\ \overline{8\ 8\ 7}\quad R$$

④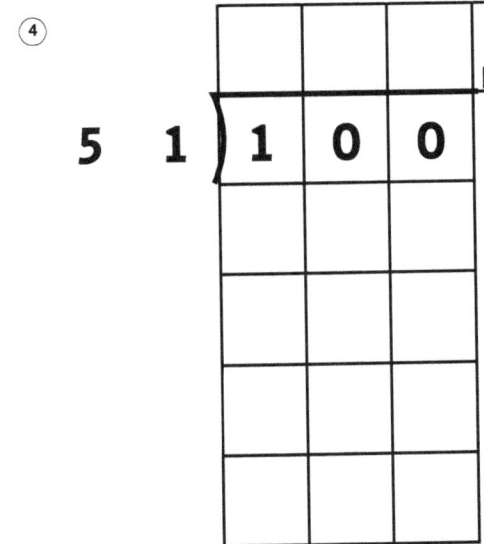

$$5\ 1\ )\ \overline{1\ 0\ 0}$$

41

3 Digits by 2 Digits

①

$$5\ 3 \overline{)6\ 4\ 5}$$ R

②

$$6\ 6 \overline{)2\ 9\ 8}$$

③

$$1\ 0 \overline{)3\ 9\ 5}$$ R

④

$$9\ 9 \overline{)8\ 2\ 1}$$

42

3 Digits by 2 Digits

①

$$1\ 3\ )\overline{6\ 3\ 2}\quad R$$

②

$$7\ 4\ )\overline{1\ 9\ 1}$$

③

$$6\ 5\ )\overline{9\ 8\ 2}\quad R$$

④

$$8\ 1\ )\overline{8\ 8\ 1}$$

43

3 Digits by 2 Digits

① 67) 3 8 3 R

② 38) 1 6 0

③ 91) 2 9 5 R

④ 94) 2 2 4

44

3 Digits by 2 Digits

① 8 1) 8 6 9 R

② 8 3) 4 2 8 |

③ 2 2) 1 1 4 R

④ 6 8) 4 9 6 |

3 Digits by 2 Digits

① 6 5)6 7 2   R

② 3 0)5 7 7

③ 9 4)3 7 2   R

④ 7 0)7 6 0

46

3 Digits by 2 Digits

① 2 8) 9 9 3 R

② 2 6) 2 7 4

③ 1 3) 4 4 5 R

④ 9 5) 2 6 1

3 Digits by 2 Digits

①
$$89{\overline{\smash{\big)}\,}}546 \text{ R}$$

②
$$11{\overline{\smash{\big)}\,}}764 \text{ }$$

③
$$32{\overline{\smash{\big)}\,}}830 \text{ R}$$

④
$$57{\overline{\smash{\big)}\,}}325 \text{ }$$

3 Digits by 2 Digits

① 6 0) 2 5 5 R

② 7 0) 4 8 3

③ 2 6) 1 4 1 R

④ 4 1) 6 1 2

49

3 Digits by 2 Digits

① 9 3) 2 7 5   R

② 8 5) 1 5 2

③ 3 0) 5 4 8   R

④ 2 2) 2 5 1

50

3 Digits by 2 Digits

① 6 1)9 2 7 R

② 4 8)5 5 0 I

③ 4 8)9 5 2 R

④ 6 4)3 7 9 I

51

3 Digits by 2 Digits

①
$$7\,5\,)\,\overline{3\,6\,8}\quad R$$

②
$$5\,0\,)\,\overline{6\,5\,7}\quad I$$

③
$$4\,4\,)\,\overline{8\,6\,0}\quad R$$

④
$$8\,3\,)\,\overline{5\,4\,3}\quad I$$

3 Digits by 2 Digits

①

$$6\ 5\ \overline{)\ 9\ 0\ 3\ }\ R$$

②

$$1\ 1\ \overline{)\ 5\ 3\ 0\ }$$

③

$$9\ 5\ \overline{)\ 1\ 3\ 2\ }\ R$$

④

$$8\ 7\ \overline{)\ 9\ 0\ 3\ }$$

53

3 Digits by 2 Digits

① 29)842 R

② 71)728 I

③ 15)258 R

④ 99)207 I

54

Section Three
Fractions

Adding Fractions

1) $\frac{3}{6} + \frac{1}{3} =$ 2) $\frac{2}{5} + \frac{3}{9} =$ 3) $\frac{1}{2} + \frac{1}{3} =$

4) $\frac{2}{9} + \frac{4}{6} =$ 5) $\frac{1}{4} + \frac{3}{6} =$ 6) $\frac{1}{10} + \frac{1}{2} =$

7) $\frac{1}{2} + \frac{7}{9} =$ 8) $\frac{7}{9} + \frac{5}{10} =$ 9) $\frac{5}{7} + \frac{1}{2} =$

10) $\frac{2}{7} + \frac{1}{5} =$ 11) $\frac{3}{5} + \frac{6}{9} =$ 12) $\frac{2}{5} + \frac{2}{10} =$

13) $\frac{7}{9} + \frac{5}{8} =$ 14) $\frac{1}{3} + \frac{1}{10} =$ 15) $\frac{8}{10} + \frac{1}{4} =$

16) $\frac{3}{10} + \frac{7}{8} =$ 17) $\frac{3}{4} + \frac{3}{6} =$ 18) $\frac{1}{5} + \frac{2}{4} =$

19) $\frac{1}{3} + \frac{7}{10} =$ 20) $\frac{3}{8} + \frac{1}{2} =$ 21) $\frac{5}{10} + \frac{4}{5} =$

22) $\frac{4}{7} + \frac{2}{5} =$ 23) $\frac{2}{10} + \frac{3}{8} =$ 24) $\frac{2}{3} + \frac{5}{6} =$

25) $\frac{6}{9} + \frac{8}{10} =$ 26) $\frac{5}{6} + \frac{3}{5} =$ 27) $\frac{5}{8} + \frac{4}{10} =$

Adding Fractions

1) $\dfrac{1}{4} + \dfrac{1}{2} =$

2) $\dfrac{8}{9} + \dfrac{3}{7} =$

3) $\dfrac{1}{3} + \dfrac{1}{7} =$

4) $\dfrac{1}{3} + \dfrac{1}{5} =$

5) $\dfrac{1}{4} + \dfrac{2}{9} =$

6) $\dfrac{4}{5} + \dfrac{4}{8} =$

7) $\dfrac{3}{9} + \dfrac{2}{4} =$

8) $\dfrac{1}{3} + \dfrac{4}{10} =$

9) $\dfrac{2}{5} + \dfrac{3}{9} =$

10) $\dfrac{2}{6} + \dfrac{1}{3} =$

11) $\dfrac{3}{4} + \dfrac{3}{6} =$

12) $\dfrac{8}{10} + \dfrac{5}{7} =$

13) $\dfrac{2}{3} + \dfrac{5}{9} =$

14) $\dfrac{3}{5} + \dfrac{2}{7} =$

15) $\dfrac{4}{6} + \dfrac{2}{10} =$

16) $\dfrac{6}{10} + \dfrac{1}{7} =$

17) $\dfrac{1}{9} + \dfrac{3}{8} =$

18) $\dfrac{6}{9} + \dfrac{1}{7} =$

19) $\dfrac{1}{9} + \dfrac{3}{8} =$

20) $\dfrac{5}{7} + \dfrac{2}{3} =$

21) $\dfrac{3}{4} + \dfrac{1}{3} =$

22) $\dfrac{1}{10} + \dfrac{2}{3} =$

23) $\dfrac{2}{4} + \dfrac{4}{5} =$

24) $\dfrac{1}{6} + \dfrac{5}{8} =$

25) $\dfrac{4}{5} + \dfrac{6}{10} =$

26) $\dfrac{1}{2} + \dfrac{1}{3} =$

27) $\dfrac{4}{9} + \dfrac{4}{5} =$

Adding Fractions

(1) $\dfrac{5}{6} + \dfrac{3}{5} =$ (2) $\dfrac{3}{8} + \dfrac{1}{3} =$ (3) $\dfrac{2}{5} + \dfrac{2}{8} =$

(4) $\dfrac{2}{3} + \dfrac{5}{9} =$ (5) $\dfrac{2}{6} + \dfrac{2}{5} =$ (6) $\dfrac{1}{2} + \dfrac{1}{3} =$

(7) $\dfrac{4}{7} + \dfrac{2}{9} =$ (8) $\dfrac{9}{10} + \dfrac{1}{2} =$ (9) $\dfrac{5}{10} + \dfrac{2}{9} =$

(10) $\dfrac{8}{10} + \dfrac{5}{7} =$ (11) $\dfrac{1}{2} + \dfrac{5}{7} =$ (12) $\dfrac{1}{2} + \dfrac{7}{10} =$

(13) $\dfrac{1}{3} + \dfrac{2}{5} =$ (14) $\dfrac{1}{2} + \dfrac{3}{4} =$ (15) $\dfrac{6}{7} + \dfrac{2}{9} =$

(16) $\dfrac{7}{8} + \dfrac{2}{9} =$ (17) $\dfrac{8}{9} + \dfrac{1}{3} =$ (18) $\dfrac{4}{6} + \dfrac{1}{2} =$

(19) $\dfrac{1}{2} + \dfrac{8}{9} =$ (20) $\dfrac{4}{9} + \dfrac{1}{8} =$ (21) $\dfrac{3}{9} + \dfrac{5}{7} =$

(22) $\dfrac{5}{8} + \dfrac{4}{10} =$ (23) $\dfrac{3}{7} + \dfrac{3}{6} =$ (24) $\dfrac{1}{2} + \dfrac{2}{3} =$

(25) $\dfrac{2}{3} + \dfrac{3}{6} =$ (26) $\dfrac{3}{4} + \dfrac{2}{9} =$ (27) $\dfrac{5}{8} + \dfrac{1}{2} =$

Adding Fractions

1) $\dfrac{5}{9} + \dfrac{1}{2} =$

2) $\dfrac{1}{3} + \dfrac{3}{4} =$

3) $\dfrac{1}{6} + \dfrac{2}{3} =$

4) $\dfrac{3}{7} + \dfrac{3}{6} =$

5) $\dfrac{2}{3} + \dfrac{3}{8} =$

6) $\dfrac{5}{6} + \dfrac{4}{9} =$

7) $\dfrac{4}{5} + \dfrac{7}{9} =$

8) $\dfrac{2}{3} + \dfrac{1}{2} =$

9) $\dfrac{2}{4} + \dfrac{6}{8} =$

10) $\dfrac{3}{6} + \dfrac{5}{8} =$

11) $\dfrac{1}{3} + \dfrac{1}{6} =$

12) $\dfrac{3}{5} + \dfrac{5}{7} =$

13) $\dfrac{4}{5} + \dfrac{2}{8} =$

14) $\dfrac{1}{2} + \dfrac{3}{5} =$

15) $\dfrac{2}{3} + \dfrac{1}{6} =$

16) $\dfrac{4}{8} + \dfrac{4}{9} =$

17) $\dfrac{1}{5} + \dfrac{4}{7} =$

18) $\dfrac{1}{10} + \dfrac{4}{9} =$

19) $\dfrac{3}{4} + \dfrac{6}{8} =$

20) $\dfrac{4}{10} + \dfrac{2}{4} =$

21) $\dfrac{1}{2} + \dfrac{1}{10} =$

22) $\dfrac{5}{6} + \dfrac{9}{10} =$

23) $\dfrac{3}{7} + \dfrac{1}{10} =$

24) $\dfrac{2}{9} + \dfrac{1}{2} =$

25) $\dfrac{8}{9} + \dfrac{2}{4} =$

26) $\dfrac{1}{4} + \dfrac{2}{6} =$

27) $\dfrac{5}{10} + \dfrac{7}{8} =$

59

Adding Fractions

1) $\dfrac{5}{7} + \dfrac{3}{4} =$ 2) $\dfrac{2}{3} + \dfrac{3}{5} =$ 3) $\dfrac{5}{6} + \dfrac{1}{4} =$

4) $\dfrac{4}{8} + \dfrac{1}{5} =$ 5) $\dfrac{1}{6} + \dfrac{3}{4} =$ 6) $\dfrac{3}{6} + \dfrac{2}{10} =$

7) $\dfrac{1}{2} + \dfrac{1}{5} =$ 8) $\dfrac{1}{4} + \dfrac{2}{7} =$ 9) $\dfrac{4}{5} + \dfrac{3}{4} =$

10) $\dfrac{2}{4} + \dfrac{4}{9} =$ 11) $\dfrac{3}{8} + \dfrac{6}{7} =$ 12) $\dfrac{8}{9} + \dfrac{2}{5} =$

13) $\dfrac{2}{3} + \dfrac{4}{10} =$ 14) $\dfrac{4}{6} + \dfrac{5}{7} =$ 15) $\dfrac{2}{3} + \dfrac{2}{8} =$

16) $\dfrac{7}{9} + \dfrac{2}{3} =$ 17) $\dfrac{2}{5} + \dfrac{4}{6} =$ 18) $\dfrac{1}{3} + \dfrac{7}{10} =$

19) $\dfrac{2}{5} + \dfrac{5}{9} =$ 20) $\dfrac{8}{10} + \dfrac{5}{8} =$ 21) $\dfrac{5}{8} + \dfrac{4}{6} =$

22) $\dfrac{9}{10} + \dfrac{6}{7} =$ 23) $\dfrac{5}{9} + \dfrac{1}{4} =$ 24) $\dfrac{2}{9} + \dfrac{4}{5} =$

25) $\dfrac{1}{3} + \dfrac{6}{8} =$ 26) $\dfrac{1}{5} + \dfrac{4}{8} =$ 27) $\dfrac{7}{8} + \dfrac{6}{7} =$

Adding Fractions

1) $\dfrac{6}{8} + \dfrac{5}{10} =$ 2) $\dfrac{4}{5} + \dfrac{1}{3} =$ 3) $\dfrac{3}{5} + \dfrac{1}{2} =$

4) $\dfrac{3}{5} + \dfrac{1}{2} =$ 5) $\dfrac{3}{4} + \dfrac{1}{8} =$ 6) $\dfrac{3}{7} + \dfrac{1}{2} =$

7) $\dfrac{4}{7} + \dfrac{6}{9} =$ 8) $\dfrac{4}{8} + \dfrac{1}{4} =$ 9) $\dfrac{6}{10} + \dfrac{5}{8} =$

10) $\dfrac{6}{7} + \dfrac{2}{5} =$ 11) $\dfrac{2}{5} + \dfrac{4}{6} =$ 12) $\dfrac{4}{9} + \dfrac{2}{8} =$

13) $\dfrac{3}{4} + \dfrac{2}{6} =$ 14) $\dfrac{2}{4} + \dfrac{2}{6} =$ 15) $\dfrac{2}{5} + \dfrac{3}{6} =$

16) $\dfrac{1}{5} + \dfrac{1}{2} =$ 17) $\dfrac{7}{9} + \dfrac{2}{8} =$ 18) $\dfrac{2}{6} + \dfrac{4}{8} =$

19) $\dfrac{5}{6} + \dfrac{3}{4} =$ 20) $\dfrac{1}{7} + \dfrac{2}{3} =$ 21) $\dfrac{1}{3} + \dfrac{3}{7} =$

22) $\dfrac{2}{3} + \dfrac{1}{4} =$ 23) $\dfrac{4}{7} + \dfrac{6}{10} =$ 24) $\dfrac{6}{7} + \dfrac{1}{2} =$

25) $\dfrac{2}{3} + \dfrac{6}{10} =$ 26) $\dfrac{2}{5} + \dfrac{1}{7} =$ 27) $\dfrac{2}{9} + \dfrac{3}{6} =$

Adding Mixed Numbers

1. $1\frac{2}{10} + 1\frac{2}{8} =$

2. $7\frac{1}{2} + 1\frac{3}{5} =$

3. $1\frac{4}{8} + 1\frac{1}{10} =$

4. $6\frac{1}{3} + 3\frac{1}{5} =$

5. $3\frac{2}{4} + 2\frac{1}{7} =$

6. $3\frac{1}{6} + 2\frac{3}{7} =$

7. $1\frac{3}{10} + 1\frac{1}{6} =$

8. $3\frac{1}{6} + 1\frac{2}{4} =$

9. $1\frac{3}{9} + 2\frac{1}{8} =$

10. $1\frac{5}{10} + 2\frac{3}{7} =$

11. $1\frac{1}{4} + 1\frac{6}{7} =$

12. $1\frac{3}{9} + 1\frac{6}{10} =$

13. $2\frac{1}{9} + 2\frac{4}{7} =$

14. $3\frac{1}{2} + 2\frac{1}{6} =$

15. $6\frac{1}{2} + 2\frac{2}{8} =$

16. $2\frac{2}{9} + 1\frac{7}{8} =$

17. $4\frac{3}{4} + 2\frac{2}{5} =$

18. $2\frac{1}{8} + 4\frac{1}{2} =$

Adding Mixed Numbers

(1) $2\frac{4}{8} + 1\frac{1}{4} =$

(2) $1\frac{4}{10} + 2\frac{2}{9} =$

(3) $2\frac{5}{6} + 2\frac{1}{5} =$

(4) $2\frac{3}{8} + 4\frac{1}{2} =$

(5) $5\frac{1}{3} + 1\frac{3}{5} =$

(6) $2\frac{1}{9} + 4\frac{2}{4} =$

(7) $1\frac{8}{10} + 1\frac{4}{5} =$

(8) $2\frac{6}{7} + 1\frac{8}{10} =$

(9) $2\frac{2}{9} + 1\frac{1}{10} =$

(10) $3\frac{1}{5} + 1\frac{6}{9} =$

(11) $3\frac{1}{2} + 1\frac{1}{4} =$

(12) $1\frac{7}{10} + 1\frac{1}{4} =$

(13) $1\frac{1}{7} + 3\frac{1}{5} =$

(14) $2\frac{1}{9} + 1\frac{3}{10} =$

(15) $1\frac{4}{10} + 2\frac{1}{6} =$

(16) $3\frac{1}{2} + 5\frac{2}{3} =$

(17) $2\frac{1}{9} + 2\frac{4}{8} =$

(18) $4\frac{1}{3} + 6\frac{1}{2} =$

Adding Mixed Numbers

① $2\frac{1}{3} + 1\frac{2}{4} =$

② $1\frac{4}{10} + 1\frac{4}{7} =$

③ $2\frac{1}{9} + 1\frac{4}{8} =$

④ $4\frac{1}{3} + 1\frac{6}{7} =$

⑤ $3\frac{2}{3} + 1\frac{3}{6} =$

⑥ $2\frac{3}{4} + 1\frac{6}{8} =$

⑦ $2\frac{1}{3} + 1\frac{2}{9} =$

⑧ $3\frac{1}{4} + 2\frac{2}{6} =$

⑨ $4\frac{3}{4} + 2\frac{3}{6} =$

⑩ $4\frac{3}{4} + 8\frac{1}{2} =$

⑪ $3\frac{1}{4} + 1\frac{8}{10} =$

⑫ $9\frac{1}{2} + 1\frac{5}{7} =$

⑬ $2\frac{3}{8} + 2\frac{1}{9} =$

⑭ $1\frac{1}{8} + 1\frac{1}{10} =$

⑮ $3\frac{2}{6} + 3\frac{2}{4} =$

⑯ $1\frac{1}{2} + 1\frac{3}{9} =$

⑰ $1\frac{2}{3} + 1\frac{5}{8} =$

⑱ $1\frac{1}{9} + 2\frac{5}{6} =$

64

Adding Mixed Numbers

(1) $1\frac{3}{10} + 2\frac{1}{2} =$

(2) $2\frac{4}{7} + 1\frac{4}{9} =$

(3) $1\frac{7}{8} + 1\frac{6}{10} =$

(4) $1\frac{2}{5} + 1\frac{4}{10} =$

(5) $1\frac{3}{8} + 2\frac{2}{9} =$

(6) $5\frac{1}{3} + 1\frac{1}{6} =$

(7) $1\frac{2}{5} + 1\frac{4}{8} =$

(8) $1\frac{7}{9} + 1\frac{2}{6} =$

(9) $1\frac{4}{9} + 1\frac{4}{10} =$

(10) $1\frac{7}{8} + 1\frac{1}{6} =$

(11) $2\frac{1}{6} + 2\frac{4}{5} =$

(12) $1\frac{4}{9} + 2\frac{1}{3} =$

(13) $2\frac{3}{8} + 1\frac{6}{10} =$

(14) $4\frac{2}{4} + 1\frac{3}{10} =$

(15) $1\frac{8}{10} + 1\frac{5}{6} =$

(16) $1\frac{9}{10} + 1\frac{5}{7} =$

(17) $2\frac{2}{5} + 7\frac{1}{2} =$

(18) $1\frac{3}{6} + 1\frac{4}{7} =$

Adding Mixed Numbers

(1) $2\frac{3}{4} + 2\frac{3}{5} =$

(2) $1\frac{8}{9} + 1\frac{1}{7} =$

(3) $2\frac{1}{6} + 2\frac{2}{3} =$

(4) $1\frac{6}{10} + 1\frac{4}{6} =$

(5) $1\frac{4}{9} + 1\frac{4}{10} =$

(6) $1\frac{1}{8} + 1\frac{3}{5} =$

(7) $2\frac{1}{7} + 1\frac{2}{8} =$

(8) $9\frac{1}{2} + 1\frac{7}{9} =$

(9) $1\frac{6}{10} + 1\frac{4}{8} =$

(10) $1\frac{8}{10} + 1\frac{2}{5} =$

(11) $2\frac{4}{8} + 1\frac{1}{10} =$

(12) $1\frac{6}{9} + 1\frac{2}{10} =$

(13) $1\frac{5}{9} + 4\frac{2}{4} =$

(14) $1\frac{5}{8} + 1\frac{3}{9} =$

(15) $1\frac{4}{6} + 1\frac{4}{7} =$

(16) $2\frac{5}{6} + 2\frac{1}{5} =$

(17) $2\frac{3}{4} + 2\frac{2}{7} =$

(18) $3\frac{1}{4} + 6\frac{2}{3} =$

Adding Mixed Numbers

1. $1\frac{2}{4} + 1\frac{3}{10} =$

2. $2\frac{1}{7} + 1\frac{7}{9} =$

3. $1\frac{3}{9} + 8\frac{1}{2} =$

4. $1\frac{5}{9} + 4\frac{2}{3} =$

5. $1\frac{2}{9} + 2\frac{1}{5} =$

6. $3\frac{1}{5} + 2\frac{5}{7} =$

7. $4\frac{2}{4} + 1\frac{1}{10} =$

8. $1\frac{6}{9} + 1\frac{4}{7} =$

9. $2\frac{1}{6} + 3\frac{1}{3} =$

10. $2\frac{1}{6} + 1\frac{1}{7} =$

11. $2\frac{2}{9} + 2\frac{5}{7} =$

12. $6\frac{1}{3} + 2\frac{4}{8} =$

13. $2\frac{1}{3} + 1\frac{4}{8} =$

14. $4\frac{1}{2} + 1\frac{5}{7} =$

15. $1\frac{7}{9} + 1\frac{1}{4} =$

16. $3\frac{4}{5} + 1\frac{7}{8} =$

17. $1\frac{2}{8} + 2\frac{4}{5} =$

18. $1\frac{5}{7} + 1\frac{8}{9} =$

Adding Mixed Numbers

1. $1\frac{3}{10} + 2\frac{3}{8} =$

2. $1\frac{9}{10} + 1\frac{1}{7} =$

3. $2\frac{1}{3} + 1\frac{1}{10} =$

4. $1\frac{7}{8} + 3\frac{1}{3} =$

5. $1\frac{5}{6} + 1\frac{2}{7} =$

6. $4\frac{1}{4} + 2\frac{5}{6} =$

7. $2\frac{3}{6} + 1\frac{2}{7} =$

8. $2\frac{2}{8} + 1\frac{5}{10} =$

9. $2\frac{4}{8} + 1\frac{6}{9} =$

10. $6\frac{1}{2} + 2\frac{2}{9} =$

11. $1\frac{1}{7} + 2\frac{1}{4} =$

12. $1\frac{1}{10} + 1\frac{1}{8} =$

13. $1\frac{6}{7} + 3\frac{4}{5} =$

14. $2\frac{1}{5} + 7\frac{1}{2} =$

15. $2\frac{1}{8} + 1\frac{8}{10} =$

16. $3\frac{2}{4} + 1\frac{4}{10} =$

17. $2\frac{3}{4} + 1\frac{8}{10} =$

18. $1\frac{2}{7} + 2\frac{1}{5} =$

Adding Mixed Numbers

1) $3\frac{2}{6} + 1\frac{3}{8} =$

2) $1\frac{8}{9} + 2\frac{2}{6} =$

3) $2\frac{2}{4} + 1\frac{2}{7} =$

4) $3\frac{1}{6} + 1\frac{5}{9} =$

5) $1\frac{1}{8} + 1\frac{9}{10} =$

6) $4\frac{1}{3} + 2\frac{6}{7} =$

7) $2\frac{2}{7} + 1\frac{1}{10} =$

8) $1\frac{7}{9} + 3\frac{4}{5} =$

9) $2\frac{2}{9} + 4\frac{2}{4} =$

10) $1\frac{1}{9} + 1\frac{2}{4} =$

11) $1\frac{3}{10} + 4\frac{1}{2} =$

12) $3\frac{2}{4} + 6\frac{1}{3} =$

13) $1\frac{7}{9} + 1\frac{1}{6} =$

14) $3\frac{2}{6} + 2\frac{1}{5} =$

15) $2\frac{5}{7} + 1\frac{2}{5} =$

16) $3\frac{3}{5} + 3\frac{2}{3} =$

17) $1\frac{5}{7} + 6\frac{1}{3} =$

18) $1\frac{8}{10} + 1\frac{1}{7} =$

69

Adding Mixed Numbers

1. $2\dfrac{4}{5} + 1\dfrac{2}{8} =$

2. $1\dfrac{7}{10} + 2\dfrac{6}{7} =$

3. $2\dfrac{4}{8} + 1\dfrac{6}{9} =$

4. $1\dfrac{6}{8} + 1\dfrac{1}{10} =$

5. $2\dfrac{1}{6} + 2\dfrac{1}{5} =$

6. $1\dfrac{3}{9} + 2\dfrac{1}{5} =$

7. $1\dfrac{1}{10} + 1\dfrac{1}{8} =$

8. $1\dfrac{8}{9} + 2\dfrac{4}{5} =$

9. $8\dfrac{1}{2} + 1\dfrac{2}{6} =$

10. $1\dfrac{2}{10} + 2\dfrac{3}{7} =$

11. $1\dfrac{3}{8} + 1\dfrac{1}{2} =$

12. $1\dfrac{4}{10} + 1\dfrac{7}{9} =$

13. $1\dfrac{1}{5} + 8\dfrac{1}{2} =$

14. $4\dfrac{2}{4} + 2\dfrac{2}{5} =$

15. $1\dfrac{1}{5} + 1\dfrac{3}{7} =$

16. $5\dfrac{1}{3} + 1\dfrac{3}{8} =$

17. $1\dfrac{3}{7} + 2\dfrac{1}{2} =$

18. $1\dfrac{6}{7} + 4\dfrac{2}{4} =$

Subtracting Fractions

1) $\frac{1}{2} - \frac{1}{6} =$

2) $\frac{3}{4} - \frac{4}{8} =$

3) $\frac{5}{6} - \frac{4}{5} =$

4) $\frac{1}{2} - \frac{2}{5} =$

5) $\frac{5}{8} - \frac{3}{10} =$

6) $\frac{5}{7} - \frac{3}{6} =$

7) $\frac{4}{7} - \frac{3}{6} =$

8) $\frac{1}{2} - \frac{2}{8} =$

9) $\frac{7}{9} - \frac{1}{2} =$

10) $\frac{9}{10} - \frac{1}{3} =$

11) $\frac{1}{2} - \frac{2}{6} =$

12) $\frac{2}{4} - \frac{2}{6} =$

13) $\frac{6}{8} - \frac{1}{6} =$

14) $\frac{1}{2} - \frac{4}{9} =$

15) $\frac{3}{9} - \frac{1}{4} =$

16) $\frac{8}{10} - \frac{4}{9} =$

17) $\frac{2}{5} - \frac{3}{8} =$

18) $\frac{6}{7} - \frac{3}{4} =$

19) $\frac{4}{9} - \frac{1}{5} =$

20) $\frac{2}{3} - \frac{1}{5} =$

21) $\frac{6}{10} - \frac{2}{6} =$

22) $\frac{8}{10} - \frac{1}{8} =$

23) $\frac{4}{10} - \frac{1}{9} =$

24) $\frac{5}{8} - \frac{1}{6} =$

25) $\frac{5}{6} - \frac{1}{4} =$

26) $\frac{6}{8} - \frac{6}{10} =$

27) $\frac{1}{2} - \frac{1}{4} =$

Subtracting Fractions

1) $\dfrac{7}{8} - \dfrac{7}{9} =$ 2) $\dfrac{5}{7} - \dfrac{1}{2} =$ 3) $\dfrac{6}{7} - \dfrac{2}{3} =$

4) $\dfrac{6}{8} - \dfrac{1}{4} =$ 5) $\dfrac{1}{2} - \dfrac{4}{9} =$ 6) $\dfrac{6}{9} - \dfrac{2}{7} =$

7) $\dfrac{3}{10} - \dfrac{1}{4} =$ 8) $\dfrac{3}{4} - \dfrac{4}{6} =$ 9) $\dfrac{7}{9} - \dfrac{2}{3} =$

10) $\dfrac{5}{8} - \dfrac{1}{2} =$ 11) $\dfrac{3}{4} - \dfrac{4}{10} =$ 12) $\dfrac{5}{8} - \dfrac{2}{5} =$

13) $\dfrac{1}{2} - \dfrac{1}{4} =$ 14) $\dfrac{6}{10} - \dfrac{1}{4} =$ 15) $\dfrac{1}{2} - \dfrac{3}{10} =$

16) $\dfrac{1}{2} - \dfrac{2}{8} =$ 17) $\dfrac{2}{7} - \dfrac{2}{9} =$ 18) $\dfrac{6}{7} - \dfrac{5}{8} =$

19) $\dfrac{1}{3} - \dfrac{1}{10} =$ 20) $\dfrac{4}{10} - \dfrac{2}{6} =$ 21) $\dfrac{1}{2} - \dfrac{1}{4} =$

22) $\dfrac{8}{9} - \dfrac{8}{10} =$ 23) $\dfrac{3}{5} - \dfrac{4}{8} =$ 24) $\dfrac{1}{2} - \dfrac{2}{6} =$

25) $\dfrac{7}{9} - \dfrac{1}{4} =$ 26) $\dfrac{5}{7} - \dfrac{2}{6} =$ 27) $\dfrac{8}{9} - \dfrac{5}{8} =$

Subtracting Fractions

1) $\dfrac{7}{9} - \dfrac{2}{5} =$

2) $\dfrac{6}{7} - \dfrac{3}{6} =$

3) $\dfrac{6}{9} - \dfrac{2}{4} =$

4) $\dfrac{2}{4} - \dfrac{2}{8} =$

5) $\dfrac{5}{8} - \dfrac{2}{4} =$

6) $\dfrac{4}{5} - \dfrac{2}{9} =$

7) $\dfrac{4}{6} - \dfrac{1}{7} =$

8) $\dfrac{4}{9} - \dfrac{1}{4} =$

9) $\dfrac{7}{8} - \dfrac{1}{2} =$

10) $\dfrac{7}{9} - \dfrac{3}{4} =$

11) $\dfrac{2}{3} - \dfrac{2}{4} =$

12) $\dfrac{5}{8} - \dfrac{2}{6} =$

13) $\dfrac{1}{2} - \dfrac{2}{5} =$

14) $\dfrac{5}{7} - \dfrac{1}{8} =$

15) $\dfrac{4}{9} - \dfrac{2}{6} =$

16) $\dfrac{4}{5} - \dfrac{2}{3} =$

17) $\dfrac{3}{7} - \dfrac{1}{6} =$

18) $\dfrac{3}{9} - \dfrac{1}{4} =$

19) $\dfrac{2}{3} - \dfrac{2}{5} =$

20) $\dfrac{8}{10} - \dfrac{4}{7} =$

21) $\dfrac{5}{6} - \dfrac{1}{3} =$

22) $\dfrac{2}{6} - \dfrac{2}{8} =$

23) $\dfrac{3}{4} - \dfrac{6}{9} =$

24) $\dfrac{4}{9} - \dfrac{1}{6} =$

25) $\dfrac{6}{10} - \dfrac{1}{9} =$

26) $\dfrac{1}{2} - \dfrac{1}{3} =$

27) $\dfrac{4}{5} - \dfrac{3}{6} =$

Subtracting Fractions

1) $\dfrac{8}{10} - \dfrac{1}{6} =$

2) $\dfrac{1}{2} - \dfrac{2}{5} =$

3) $\dfrac{1}{2} - \dfrac{1}{8} =$

4) $\dfrac{6}{7} - \dfrac{1}{2} =$

5) $\dfrac{2}{3} - \dfrac{5}{10} =$

6) $\dfrac{1}{2} - \dfrac{1}{6} =$

7) $\dfrac{2}{3} - \dfrac{1}{4} =$

8) $\dfrac{8}{10} - \dfrac{1}{2} =$

9) $\dfrac{1}{2} - \dfrac{1}{7} =$

10) $\dfrac{2}{3} - \dfrac{5}{9} =$

11) $\dfrac{9}{10} - \dfrac{1}{4} =$

12) $\dfrac{4}{5} - \dfrac{1}{2} =$

13) $\dfrac{8}{10} - \dfrac{1}{2} =$

14) $\dfrac{3}{9} - \dfrac{1}{5} =$

15) $\dfrac{2}{3} - \dfrac{1}{2} =$

16) $\dfrac{5}{6} - \dfrac{1}{7} =$

17) $\dfrac{3}{6} - \dfrac{1}{4} =$

18) $\dfrac{9}{10} - \dfrac{2}{3} =$

19) $\dfrac{7}{8} - \dfrac{1}{7} =$

20) $\dfrac{9}{10} - \dfrac{4}{5} =$

21) $\dfrac{5}{10} - \dfrac{1}{4} =$

22) $\dfrac{2}{4} - \dfrac{1}{5} =$

23) $\dfrac{7}{8} - \dfrac{1}{4} =$

24) $\dfrac{1}{3} - \dfrac{1}{6} =$

25) $\dfrac{3}{5} - \dfrac{1}{4} =$

26) $\dfrac{8}{10} - \dfrac{5}{9} =$

27) $\dfrac{3}{4} - \dfrac{2}{7} =$

Subtracting Fractions

1) $\dfrac{4}{5} - \dfrac{1}{2} =$

2) $\dfrac{1}{2} - \dfrac{1}{4} =$

3) $\dfrac{1}{2} - \dfrac{1}{4} =$

4) $\dfrac{1}{2} - \dfrac{1}{3} =$

5) $\dfrac{2}{3} - \dfrac{4}{7} =$

6) $\dfrac{4}{5} - \dfrac{3}{4} =$

7) $\dfrac{4}{5} - \dfrac{2}{10} =$

8) $\dfrac{5}{8} - \dfrac{1}{4} =$

9) $\dfrac{6}{7} - \dfrac{8}{10} =$

10) $\dfrac{8}{9} - \dfrac{7}{8} =$

11) $\dfrac{3}{4} - \dfrac{4}{6} =$

12) $\dfrac{4}{5} - \dfrac{3}{6} =$

13) $\dfrac{4}{5} - \dfrac{1}{10} =$

14) $\dfrac{5}{9} - \dfrac{2}{4} =$

15) $\dfrac{2}{3} - \dfrac{1}{9} =$

16) $\dfrac{3}{4} - \dfrac{1}{5} =$

17) $\dfrac{8}{9} - \dfrac{3}{8} =$

18) $\dfrac{2}{3} - \dfrac{1}{4} =$

19) $\dfrac{1}{2} - \dfrac{1}{7} =$

20) $\dfrac{3}{4} - \dfrac{2}{3} =$

21) $\dfrac{2}{3} - \dfrac{1}{7} =$

22) $\dfrac{3}{4} - \dfrac{5}{8} =$

23) $\dfrac{5}{6} - \dfrac{1}{5} =$

24) $\dfrac{2}{3} - \dfrac{1}{9} =$

25) $\dfrac{4}{7} - \dfrac{1}{2} =$

26) $\dfrac{3}{5} - \dfrac{3}{6} =$

27) $\dfrac{2}{3} - \dfrac{1}{2} =$

Subtracting Fractions

1) $\dfrac{3}{4} - \dfrac{1}{2} =$

2) $\dfrac{5}{7} - \dfrac{2}{5} =$

3) $\dfrac{1}{2} - \dfrac{1}{4} =$

4) $\dfrac{3}{4} - \dfrac{1}{3} =$

5) $\dfrac{1}{2} - \dfrac{1}{6} =$

6) $\dfrac{5}{6} - \dfrac{1}{3} =$

7) $\dfrac{1}{2} - \dfrac{1}{5} =$

8) $\dfrac{2}{3} - \dfrac{2}{5} =$

9) $\dfrac{5}{9} - \dfrac{1}{4} =$

10) $\dfrac{3}{7} - \dfrac{1}{3} =$

11) $\dfrac{3}{10} - \dfrac{2}{9} =$

12) $\dfrac{1}{2} - \dfrac{1}{4} =$

13) $\dfrac{1}{2} - \dfrac{3}{7} =$

14) $\dfrac{3}{4} - \dfrac{1}{6} =$

15) $\dfrac{3}{5} - \dfrac{2}{10} =$

16) $\dfrac{2}{5} - \dfrac{3}{8} =$

17) $\dfrac{2}{3} - \dfrac{2}{5} =$

18) $\dfrac{6}{10} - \dfrac{4}{7} =$

19) $\dfrac{3}{5} - \dfrac{1}{3} =$

20) $\dfrac{5}{8} - \dfrac{1}{6} =$

21) $\dfrac{3}{5} - \dfrac{1}{4} =$

22) $\dfrac{2}{3} - \dfrac{5}{10} =$

23) $\dfrac{1}{2} - \dfrac{1}{7} =$

24) $\dfrac{3}{7} - \dfrac{3}{9} =$

25) $\dfrac{1}{2} - \dfrac{1}{8} =$

26) $\dfrac{6}{8} - \dfrac{4}{7} =$

27) $\dfrac{2}{3} - \dfrac{2}{6} =$

Subtracting Mixed Numbers

1) $3\frac{1}{3} - 1\frac{2}{8} =$

2) $2\frac{3}{7} - 1\frac{8}{9} =$

3) $3\frac{2}{5} - 2\frac{1}{8} =$

4) $4\frac{1}{2} - 1\frac{1}{8} =$

5) $4\frac{1}{4} - 1\frac{3}{9} =$

6) $4\frac{1}{3} - 2\frac{5}{7} =$

7) $1\frac{3}{4} - 1\frac{1}{10} =$

8) $1\frac{3}{4} - 1\frac{2}{5} =$

9) $1\frac{4}{7} - 1\frac{5}{9} =$

10) $2\frac{3}{4} - 1\frac{4}{9} =$

11) $2\frac{1}{6} - 1\frac{4}{10} =$

12) $3\frac{3}{4} - 2\frac{3}{5} =$

13) $2\frac{1}{2} - 1\frac{3}{9} =$

14) $3\frac{2}{4} - 1\frac{1}{10} =$

15) $1\frac{4}{7} - 1\frac{3}{10} =$

16) $1\frac{5}{8} - 1\frac{6}{10} =$

17) $1\frac{2}{6} - 1\frac{2}{7} =$

18) $2\frac{2}{4} - 1\frac{4}{10} =$

Subtracting Mixed Numbers

1) $4\frac{2}{3} - 1\frac{3}{9} =$

2) $2\frac{2}{6} - 1\frac{8}{9} =$

3) $1\frac{8}{9} - 1\frac{2}{4} =$

4) $4\frac{1}{3} - 1\frac{5}{8} =$

5) $3\frac{2}{4} - 1\frac{6}{8} =$

6) $4\frac{2}{4} - 4\frac{1}{3} =$

7) $5\frac{2}{3} - 3\frac{2}{5} =$

8) $3\frac{1}{5} - 1\frac{8}{10} =$

9) $6\frac{2}{3} - 2\frac{1}{7} =$

10) $3\frac{2}{5} - 2\frac{1}{7} =$

11) $2\frac{2}{8} - 1\frac{1}{3} =$

12) $3\frac{4}{5} - 1\frac{3}{4} =$

13) $8\frac{1}{2} - 1\frac{7}{8} =$

14) $3\frac{2}{6} - 2\frac{2}{9} =$

15) $1\frac{4}{10} - 1\frac{3}{9} =$

16) $7\frac{1}{2} - 1\frac{3}{10} =$

17) $2\frac{1}{2} - 1\frac{4}{7} =$

18) $2\frac{5}{7} - 2\frac{2}{5} =$

Subtracting Mixed Numbers

1) $3\frac{1}{5} - 2\frac{4}{7} =$

2) $3\frac{1}{3} - 1\frac{4}{5} =$

3) $2\frac{2}{9} - 1\frac{1}{4} =$

4) $2\frac{4}{7} - 1\frac{5}{8} =$

5) $1\frac{8}{9} - 1\frac{2}{7} =$

6) $6\frac{2}{3} - 1\frac{4}{10} =$

7) $3\frac{1}{6} - 1\frac{7}{9} =$

8) $1\frac{4}{9} - 1\frac{1}{10} =$

9) $4\frac{1}{4} - 1\frac{5}{9} =$

10) $2\frac{4}{6} - 2\frac{2}{5} =$

11) $1\frac{6}{7} - 1\frac{4}{8} =$

12) $3\frac{1}{2} - 2\frac{3}{8} =$

13) $3\frac{1}{4} - 1\frac{8}{10} =$

14) $3\frac{3}{5} - 2\frac{3}{8} =$

15) $8\frac{1}{2} - 1\frac{3}{10} =$

16) $1\frac{2}{7} - 1\frac{1}{8} =$

17) $2\frac{1}{2} - 2\frac{2}{7} =$

18) $2\frac{4}{8} - 1\frac{9}{10} =$

79

Subtracting Mixed Numbers

1) $8\dfrac{1}{2} - 1\dfrac{6}{7} =$

2) $4\dfrac{3}{4} - 2\dfrac{2}{9} =$

3) $3\dfrac{1}{2} - 1\dfrac{3}{4} =$

4) $2\dfrac{5}{6} - 2\dfrac{1}{8} =$

5) $3\dfrac{2}{6} - 1\dfrac{5}{9} =$

6) $1\dfrac{8}{10} - 1\dfrac{2}{7} =$

7) $1\dfrac{8}{9} - 1\dfrac{2}{3} =$

8) $2\dfrac{5}{7} - 1\dfrac{1}{8} =$

9) $1\dfrac{5}{8} - 1\dfrac{5}{9} =$

10) $2\dfrac{1}{7} - 1\dfrac{3}{9} =$

11) $2\dfrac{2}{8} - 1\dfrac{1}{5} =$

12) $2\dfrac{5}{6} - 1\dfrac{1}{10} =$

13) $2\dfrac{1}{6} - 1\dfrac{8}{10} =$

14) $5\dfrac{2}{3} - 2\dfrac{5}{6} =$

15) $3\dfrac{1}{5} - 2\dfrac{1}{6} =$

16) $5\dfrac{1}{2} - 1\dfrac{2}{8} =$

17) $2\dfrac{1}{7} - 1\dfrac{5}{9} =$

18) $2\dfrac{6}{7} - 1\dfrac{5}{6} =$

Subtracting Mixed Numbers

1) $1\frac{5}{8} - 1\frac{1}{2} =$

2) $1\frac{7}{9} - 1\frac{2}{8} =$

3) $3\frac{3}{4} - 1\frac{6}{9} =$

4) $5\frac{1}{3} - 1\frac{2}{10} =$

5) $3\frac{2}{6} - 1\frac{9}{10} =$

6) $2\frac{3}{4} - 1\frac{2}{5} =$

7) $2\frac{3}{7} - 2\frac{1}{8} =$

8) $2\frac{1}{3} - 1\frac{1}{2} =$

9) $2\frac{5}{7} - 2\frac{4}{8} =$

10) $3\frac{2}{5} - 1\frac{2}{3} =$

11) $1\frac{4}{10} - 1\frac{2}{8} =$

12) $8\frac{1}{2} - 4\frac{1}{4} =$

13) $1\frac{8}{9} - 1\frac{1}{7} =$

14) $9\frac{1}{2} - 1\frac{2}{10} =$

15) $5\frac{1}{3} - 1\frac{6}{10} =$

16) $2\frac{2}{6} - 1\frac{2}{7} =$

17) $2\frac{1}{5} - 1\frac{2}{8} =$

18) $2\frac{1}{7} - 1\frac{5}{10} =$

Subtracting Mixed Numbers

1) $3\frac{1}{6} - 2\frac{3}{7} =$

2) $3\frac{2}{5} - 1\frac{1}{3} =$

3) $2\frac{3}{6} - 1\frac{4}{9} =$

4) $9\frac{1}{2} - 2\frac{5}{6} =$

5) $1\frac{5}{9} - 1\frac{3}{7} =$

6) $2\frac{2}{9} - 1\frac{4}{8} =$

7) $1\frac{7}{9} - 1\frac{4}{7} =$

8) $3\frac{1}{5} - 1\frac{3}{7} =$

9) $5\frac{1}{3} - 1\frac{2}{5} =$

10) $1\frac{7}{9} - 1\frac{2}{4} =$

11) $2\frac{2}{4} - 1\frac{6}{10} =$

12) $2\frac{3}{5} - 1\frac{4}{10} =$

13) $3\frac{3}{4} - 1\frac{3}{9} =$

14) $5\frac{2}{3} - 2\frac{1}{2} =$

15) $2\frac{1}{3} - 1\frac{2}{10} =$

16) $2\frac{2}{5} - 2\frac{2}{8} =$

17) $2\frac{6}{7} - 1\frac{6}{10} =$

18) $2\frac{1}{2} - 1\frac{3}{4} =$

Subtracting Mixed Numbers

1) $2\frac{4}{5} - 1\frac{2}{3} =$

2) $3\frac{2}{5} - 2\frac{2}{6} =$

3) $1\frac{8}{9} - 1\frac{5}{8} =$

4) $1\frac{5}{7} - 1\frac{3}{10} =$

5) $3\frac{2}{6} - 1\frac{1}{2} =$

6) $4\frac{1}{3} - 1\frac{2}{10} =$

7) $4\frac{2}{3} - 1\frac{4}{7} =$

8) $2\frac{5}{7} - 1\frac{1}{9} =$

9) $2\frac{3}{5} - 1\frac{1}{2} =$

10) $2\frac{3}{8} - 2\frac{1}{9} =$

11) $4\frac{1}{4} - 3\frac{2}{5} =$

12) $4\frac{2}{3} - 2\frac{3}{4} =$

13) $1\frac{2}{6} - 1\frac{2}{8} =$

14) $4\frac{1}{4} - 2\frac{2}{9} =$

15) $3\frac{3}{4} - 1\frac{8}{10} =$

16) $1\frac{8}{9} - 1\frac{2}{10} =$

17) $1\frac{3}{10} - 1\frac{2}{8} =$

18) $1\frac{3}{9} - 1\frac{2}{10} =$

83

Subtracting Mixed Numbers

1) $1\frac{4}{6} - 1\frac{1}{7} =$

2) $4\frac{1}{4} - 1\frac{4}{6} =$

3) $1\frac{5}{10} - 1\frac{3}{9} =$

4) $5\frac{1}{3} - 1\frac{1}{10} =$

5) $7\frac{1}{2} - 4\frac{2}{3} =$

6) $3\frac{2}{4} - 1\frac{7}{8} =$

7) $3\frac{4}{5} - 1\frac{1}{2} =$

8) $2\frac{5}{7} - 1\frac{3}{4} =$

9) $2\frac{2}{5} - 1\frac{6}{8} =$

10) $6\frac{1}{3} - 1\frac{4}{8} =$

11) $4\frac{3}{4} - 1\frac{3}{9} =$

12) $4\frac{3}{4} - 1\frac{4}{6} =$

13) $3\frac{2}{3} - 2\frac{2}{9} =$

14) $6\frac{2}{3} - 2\frac{3}{8} =$

15) $2\frac{1}{6} - 1\frac{9}{10} =$

16) $5\frac{2}{3} - 3\frac{1}{5} =$

17) $1\frac{4}{5} - 1\frac{4}{9} =$

18) $3\frac{3}{4} - 1\frac{7}{10} =$

Subtracting Mixed Numbers

1) $1\frac{7}{10} - 1\frac{2}{8} =$

2) $4\frac{3}{4} - 2\frac{1}{2} =$

3) $3\frac{4}{5} - 1\frac{9}{10} =$

4) $2\frac{3}{5} - 1\frac{3}{4} =$

5) $1\frac{3}{5} - 1\frac{1}{8} =$

6) $6\frac{2}{3} - 2\frac{5}{6} =$

7) $2\frac{1}{3} - 1\frac{4}{8} =$

8) $3\frac{1}{4} - 2\frac{2}{9} =$

9) $2\frac{4}{5} - 2\frac{1}{6} =$

10) $3\frac{1}{5} - 1\frac{1}{7} =$

11) $5\frac{1}{2} - 2\frac{2}{5} =$

12) $2\frac{4}{6} - 1\frac{1}{3} =$

13) $6\frac{2}{3} - 3\frac{2}{6} =$

14) $5\frac{1}{2} - 3\frac{1}{6} =$

15) $4\frac{1}{3} - 1\frac{7}{9} =$

16) $7\frac{1}{2} - 2\frac{1}{8} =$

17) $3\frac{2}{4} - 1\frac{3}{10} =$

18) $1\frac{5}{8} - 1\frac{1}{9} =$

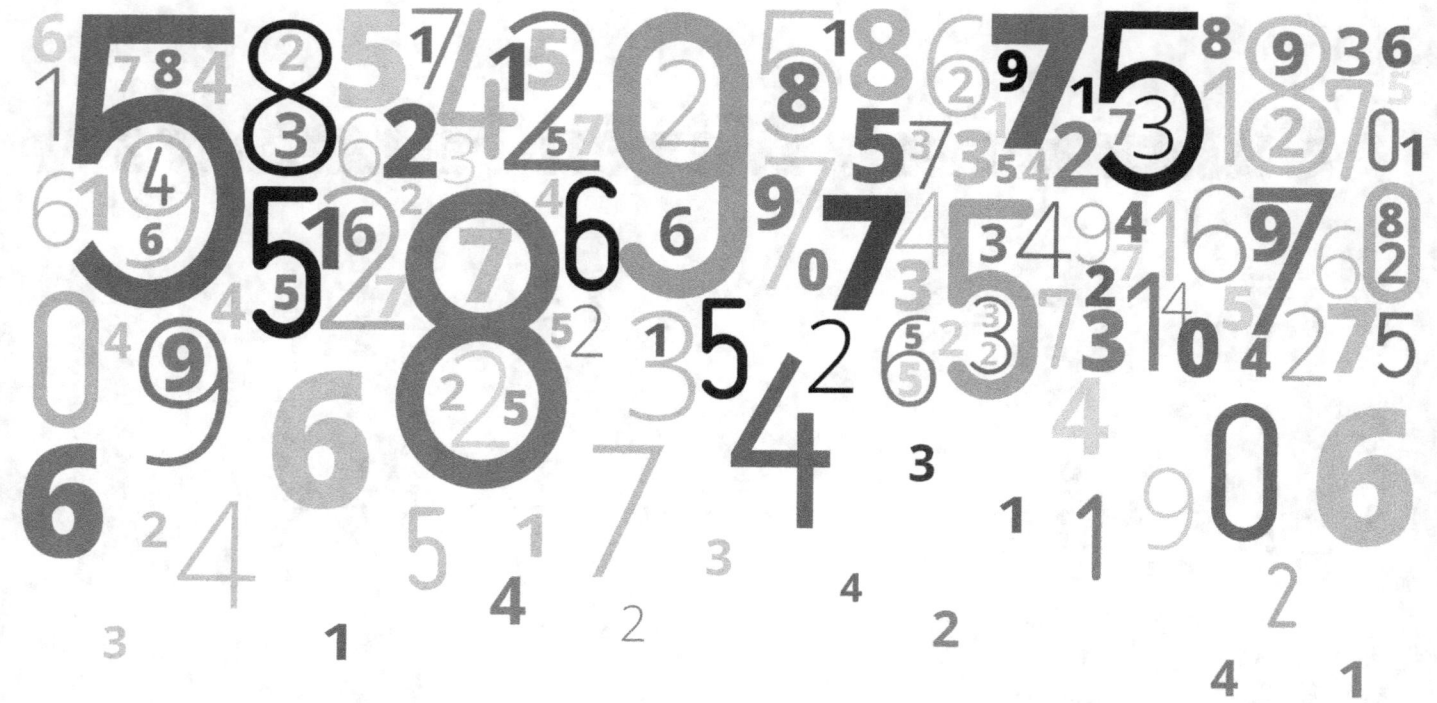

Section Four
Decimals

Adding Decimals

①   9.063
  + 5.043

②   3.175
  + 4.855

③   1.886
  + 1.895

④   1.207
  + 3.445

⑤   0.911
  + 6.998

⑥   3.933
  + 9.223

⑦   3.464
  + 4.133

⑧   6.793
  + 4.552

⑨   0.991
  + 7.193

⑩   9.114
  + 7.769

⑪   5.831
  + 8.160

⑫   5.705
  + 4.925

⑬   8.391
  + 7.624

⑭   6.689
  + 0.038

⑮   8.786
  + 4.021

⑯   4.649
  + 0.841

⑰   7.418
  + 5.248

⑱   0.805
  + 6.785

⑲   3.238
  + 1.435

⑳   9.412
  + 8.573

㉑   5.609
  + 2.417

㉒   7.501
  + 6.465

㉓   5.415
  + 6.061

㉔   4.048
  + 0.594

㉕   1.397
  + 2.086

㉖   8.715
  + 0.046

㉗   1.340
  + 0.725

㉘   0.939
  + 8.434

Adding Decimals

1)   7.480
  + 6.401

2)   9.096
  + 4.164

3)   6.810
  + 6.193

4)   5.869
  + 0.428

5)   6.009
  + 1.891

6)   0.085
  + 3.142

7)   1.516
  + 8.833

8)   6.947
  + 5.190

9)   6.010
  + 3.650

10)   5.690
  + 1.393

11)   3.747
  + 9.964

12)   1.909
  + 6.447

13)   8.586
  + 0.142

14)   5.210
  + 3.451

15)   5.734
  + 2.447

16)   3.227
  + 1.929

17)   2.049
  + 7.599

18)   5.957
  + 6.746

19)   1.817
  + 4.871

20)   9.533
  + 2.407

21)   6.934
  + 2.280

22)   4.324
  + 9.029

23)   3.545
  + 2.238

24)   9.715
  + 5.429

25)   1.162
  + 0.727

26)   8.727
  + 7.512

27)   7.106
  + 2.158

28)   7.043
  + 3.919

Adding Decimals

1)   7.088
 + 9.428

2)   2.782
 + 6.601

3)   0.960
 + 8.242

4)   5.793
 + 5.004

5)   9.258
 + 3.803

6)   9.846
 + 0.430

7)   9.090
 + 4.114

8)   7.620
 + 1.560

9)   6.240
 + 3.234

10)   0.629
 + 4.321

11)   8.081
 + 7.101

12)   6.590
 + 6.543

13)   4.485
 + 5.277

14)   0.627
 + 3.438

15)   6.617
 + 4.928

16)   6.458
 + 3.344

17)   7.660
 + 4.408

18)   3.361
 + 1.643

19)   9.611
 + 1.205

20)   7.958
 + 3.154

21)   0.484
 + 4.348

22)   6.288
 + 5.189

23)   9.028
 + 7.393

24)   3.282
 + 5.977

25)   9.697
 + 2.439

26)   3.572
 + 9.753

27)   3.127
 + 5.847

28)   7.164
 + 1.437

Adding Decimals

①   7.464
  +   8.115

②   7.582
  +   8.073

③   4.456
  +   2.713

④   5.366
  +   2.217

⑤   2.264
  +   2.937

⑥   2.807
  +   2.878

⑦   2.774
  +   0.740

⑧   4.378
  +   6.465

⑨   7.925
  +   1.100

⑩   2.842
  +   6.546

⑪   2.661
  +   2.323

⑫   1.679
  +   7.163

⑬   8.531
  +   1.577

⑭   3.833
  +   3.615

⑮   6.317
  +   9.876

⑯   1.216
  +   7.226

⑰   1.996
  +   3.969

⑱   7.027
  +   6.978

⑲   8.981
  +   4.309

⑳   4.148
  +   3.672

㉑   4.899
  +   1.707

㉒   9.094
  +   8.051

㉓   6.608
  +   3.956

㉔   4.735
  +   6.865

㉕   8.664
  +   6.418

㉖   0.803
  +   2.682

㉗   8.969
  +   2.449

㉘   3.207
  +   0.257

Adding Decimals

1)   7.679
+ 3.179

2)   6.344
+ 6.127

3)   0.839
+ 2.970

4)   1.161
+ 5.668

5)   7.197
+ 4.939

6)   5.449
+ 8.755

7)   6.216
+ 4.205

8)   1.137
+ 3.982

9)   6.611
+ 5.037

10)   7.927
+ 9.275

11)   2.987
+ 7.503

12)   9.741
+ 2.553

13)   7.828
+ 7.586

14)   7.734
+ 8.832

15)   6.834
+ 7.548

16)   4.300
+ 8.740

17)   8.881
+ 5.805

18)   1.745
+ 1.920

19)   6.607
+ 2.375

20)   4.819
+ 0.386

21)   0.394
+ 5.255

22)   1.544
+ 0.137

23)   4.783
+ 4.720

24)   5.107
+ 2.316

25)   8.737
+ 2.520

26)   3.495
+ 4.777

27)   7.102
+ 7.550

28)   5.279
+ 6.368

Adding Decimals

①   9.471
 + 2.443
_____

②   3.584
 + 1.810
_____

③   5.033
 + 9.112
_____

④   5.916
 + 8.515
_____

⑤   9.393
 + 7.294
_____

⑥   4.532
 + 4.185
_____

⑦   9.308
 + 2.916
_____

⑧   5.597
 + 5.487
_____

⑨   3.890
 + 1.312
_____

⑩   8.047
 + 0.572
_____

⑪   6.965
 + 1.987
_____

⑫   2.041
 + 1.123
_____

⑬   2.555
 + 0.423
_____

⑭   2.233
 + 1.368
_____

⑮   2.311
 + 7.250
_____

⑯   5.790
 + 6.918
_____

⑰   3.835
 + 7.271
_____

⑱   4.985
 + 4.528
_____

⑲   0.530
 + 7.024
_____

⑳   2.519
 + 5.733
_____

㉑   4.341
 + 4.109
_____

㉒   1.920
 + 8.245
_____

㉓   3.312
 + 9.054
_____

㉔   2.178
 + 1.011
_____

㉕   5.415
 + 3.625
_____

㉖   8.614
 + 7.795
_____

㉗   4.611
 + 8.030
_____

㉘   0.849
 + 7.326
_____

Subtracting Decimals

(1)	4.230 − 2.233	(2)	8.434 − 0.650	(3)	1.383 − 0.525	(4)	7.362 − 6.777
(5)	7.856 − 3.301	(6)	8.944 − 8.542	(7)	8.093 − 3.687	(8)	8.736 − 2.670
(9)	7.424 − 5.139	(10)	9.129 − 5.940	(11)	4.818 − 4.095	(12)	3.934 − 2.298
(13)	6.787 − 4.916	(14)	5.579 − 0.881	(15)	4.555 − 3.382	(16)	5.864 − 1.152
(17)	9.787 − 0.003	(18)	6.773 − 1.388	(19)	9.496 − 5.721	(20)	4.451 − 2.090
(21)	8.567 − 3.258	(22)	4.801 − 2.349	(23)	8.462 − 6.905	(24)	5.142 − 0.357
(25)	9.374 − 8.385	(26)	6.645 − 2.656	(27)	5.936 − 0.582	(28)	5.552 − 0.574

Subtracting Decimals

1)   8.909
   − 1.726
   ─────

2)   5.848
   − 4.170
   ─────

3)   5.015
   − 2.165
   ─────

4)   3.838
   − 3.566
   ─────

5)   3.891
   − 3.581
   ─────

6)   5.043
   − 1.809
   ─────

7)   5.094
   − 1.613
   ─────

8)   9.139
   − 0.611
   ─────

9)   3.447
   − 3.329
   ─────

10)   9.020
    − 0.144
    ─────

11)   8.850
    − 3.434
    ─────

12)   3.979
    − 0.360
    ─────

13)   5.677
    − 4.102
    ─────

14)   9.752
    − 1.588
    ─────

15)   6.053
    − 3.311
    ─────

16)   3.546
    − 0.586
    ─────

17)   7.208
    − 6.396
    ─────

18)   8.088
    − 4.007
    ─────

19)   7.819
    − 0.185
    ─────

20)   8.331
    − 2.653
    ─────

21)   7.381
    − 0.145
    ─────

22)   8.687
    − 4.125
    ─────

23)   9.012
    − 8.609
    ─────

24)   2.725
    − 2.036
    ─────

25)   1.795
    − 1.001
    ─────

26)   7.839
    − 1.720
    ─────

27)   6.545
    − 0.034
    ─────

28)   5.323
    − 1.723
    ─────

Subtracting Decimals

①   2.389
   − 1.182
_____

②   3.884
   − 3.546
_____

③   3.861
   − 1.569
_____

④   9.694
   − 2.556
_____

⑤   7.139
   − 3.318
_____

⑥   5.947
   − 3.356
_____

⑦   7.792
   − 0.314
_____

⑧   4.650
   − 2.268
_____

⑨   9.964
   − 3.366
_____

⑩   4.974
   − 2.754
_____

⑪   4.173
   − 2.396
_____

⑫   8.719
   − 4.522
_____

⑬   3.882
   − 1.775
_____

⑭   7.932
   − 2.830
_____

⑮   8.375
   − 5.686
_____

⑯   8.463
   − 2.366
_____

⑰   7.462
   − 1.536
_____

⑱   7.221
   − 5.949
_____

⑲   5.670
   − 5.217
_____

⑳   8.610
   − 7.863
_____

㉑   3.630
   − 0.285
_____

㉒   9.897
   − 0.635
_____

㉓   6.933
   − 1.748
_____

㉔   0.716
   − 0.212
_____

㉕   8.822
   − 2.084
_____

㉖   9.936
   − 7.457
_____

㉗   7.236
   − 5.750
_____

㉘   8.832
   − 0.115
_____

Subtracting Decimals

①   7.961
  − 6.469

②   8.413
  − 2.583

③   5.810
  − 2.654

④   3.426
  − 2.801

⑤   8.862
  − 3.320

⑥   7.877
  − 1.063

⑦   7.671
  − 7.544

⑧   9.629
  − 7.988

⑨   8.528
  − 2.612

⑩   9.668
  − 5.377

⑪   8.378
  − 4.037

⑫   3.912
  − 2.492

⑬   8.152
  − 1.899

⑭   8.954
  − 8.309

⑮   9.375
  − 2.936

⑯   9.654
  − 7.300

⑰   9.461
  − 3.539

⑱   1.881
  − 0.225

⑲   8.724
  − 2.418

⑳   6.485
  − 3.822

㉑   4.811
  − 2.047

㉒   9.047
  − 3.307

㉓   6.674
  − 1.477

㉔   7.118
  − 7.011

㉕   7.543
  − 3.161

㉖   6.924
  − 0.945

㉗   8.266
  − 6.789

㉘   6.341
  − 4.226

Subtracting Decimals

$\begin{array}{r} 7.143 \\ -\ 6.910 \\ \hline \end{array}$ (1)

$\begin{array}{r} 6.282 \\ -\ 2.124 \\ \hline \end{array}$ (2)

$\begin{array}{r} 7.936 \\ -\ 4.589 \\ \hline \end{array}$ (3)

$\begin{array}{r} 7.347 \\ -\ 3.414 \\ \hline \end{array}$ (4)

(5) $\begin{array}{r} 7.236 \\ -\ 3.914 \\ \hline \end{array}$

(6) $\begin{array}{r} 8.032 \\ -\ 2.843 \\ \hline \end{array}$

(7) $\begin{array}{r} 5.062 \\ -\ 4.746 \\ \hline \end{array}$

(8) $\begin{array}{r} 6.882 \\ -\ 6.786 \\ \hline \end{array}$

(9) $\begin{array}{r} 2.813 \\ -\ 1.367 \\ \hline \end{array}$

(10) $\begin{array}{r} 8.841 \\ -\ 4.106 \\ \hline \end{array}$

(11) $\begin{array}{r} 8.121 \\ -\ 4.695 \\ \hline \end{array}$

(12) $\begin{array}{r} 7.893 \\ -\ 4.850 \\ \hline \end{array}$

(13) $\begin{array}{r} 7.057 \\ -\ 4.701 \\ \hline \end{array}$

(14) $\begin{array}{r} 7.642 \\ -\ 0.375 \\ \hline \end{array}$

(15) $\begin{array}{r} 8.848 \\ -\ 8.099 \\ \hline \end{array}$

(16) $\begin{array}{r} 4.971 \\ -\ 0.149 \\ \hline \end{array}$

(17) $\begin{array}{r} 8.703 \\ -\ 7.300 \\ \hline \end{array}$

(18) $\begin{array}{r} 7.070 \\ -\ 1.971 \\ \hline \end{array}$

(19) $\begin{array}{r} 5.779 \\ -\ 4.170 \\ \hline \end{array}$

(20) $\begin{array}{r} 9.204 \\ -\ 0.672 \\ \hline \end{array}$

(21) $\begin{array}{r} 7.095 \\ -\ 4.437 \\ \hline \end{array}$

(22) $\begin{array}{r} 5.147 \\ -\ 5.088 \\ \hline \end{array}$

(23) $\begin{array}{r} 5.914 \\ -\ 4.379 \\ \hline \end{array}$

(24) $\begin{array}{r} 5.719 \\ -\ 2.840 \\ \hline \end{array}$

(25) $\begin{array}{r} 6.588 \\ -\ 4.614 \\ \hline \end{array}$

(26) $\begin{array}{r} 8.659 \\ -\ 7.469 \\ \hline \end{array}$

(27) $\begin{array}{r} 2.097 \\ -\ 0.325 \\ \hline \end{array}$

(28) $\begin{array}{r} 2.555 \\ -\ 1.509 \\ \hline \end{array}$

Subtracting Decimals

(1) 8.189
   − 0.703

(2) 6.698
   − 4.191

(3) 9.365
   − 7.426

(4) 2.124
   − 1.597

(5) 1.922
   − 0.432

(6) 4.501
   − 1.997

(7) 8.765
   − 6.940

(8) 4.462
   − 2.036

(9) 9.704
   − 6.482

(10) 9.467
    − 3.457

(11) 4.074
    − 3.456

(12) 8.668
    − 5.831

(13) 5.868
    − 2.009

(14) 4.982
    − 2.518

(15) 5.362
    − 0.696

(16) 4.616
    − 2.782

(17) 5.919
    − 3.145

(18) 3.146
    − 1.067

(19) 8.940
    − 3.566

(20) 6.993
    − 0.417

(21) 8.397
    − 1.961

(22) 8.484
    − 7.268

(23) 9.356
    − 0.276

(24) 3.505
    − 2.955

(25) 8.682
    − 6.973

(26) 5.134
    − 3.914

(27) 5.897
    − 0.210

(28) 7.575
    − 2.990

Multiplying Decimals

(1) 1.4
   × 5.5
   ———

(2) 4.3
   × 9.2
   ———

(3) 7.0
   × 4.5
   ———

(4) 8.3
   × 1.3
   ———

(5) 4.6
   × 8.4
   ———

(6) 0.5
   × 7.2
   ———

(7) 5.1
   × 2.7
   ———

(8) 8.7
   × 3.9
   ———

(9) 8.4
   × 6.4
   ———

(10) 8.3
   × 4.2
   ———

(11) 9.1
   × 5.9
   ———

(12) 4.4
   × 1.7
   ———

(13) 1.2
   × 1.3
   ———

(14) 6.3
   × 9.6
   ———

(15) 0.3
   × 5.6
   ———

(16) 1.8
   × 4.5
   ———

(17) 7.2
   × 0.6
   ———

(18) 4.4
   × 5.9
   ———

(19) 9.7
   × 4.0
   ———

(20) 0.3
   × 3.8
   ———

Multiplying Decimals

①   4.7
  × 9.4
_____

②   2.7
  × 2.4
_____

③   5.0
  × 3.4
_____

④   8.0
  × 8.7
_____

⑤   1.2
  × 5.4
_____

⑥   5.2
  × 8.3
_____

⑦   9.5
  × 9.3
_____

⑧   9.6
  × 6.5
_____

⑨   2.4
  × 8.7
_____

⑩   1.1
  × 4.8
_____

⑪   0.1
  × 5.1
_____

⑫   5.2
  × 9.3
_____

⑬   2.6
  × 4.7
_____

⑭   1.6
  × 4.0
_____

⑮   8.4
  × 4.9
_____

⑯   6.9
  × 0.3
_____

⑰   2.8
  × 0.6
_____

⑱   9.5
  × 2.8
_____

⑲   7.0
  × 2.0
_____

⑳   5.8
  × 1.9
_____

Multiplying Decimals

(1)   7.5
 × 6.8

(2)   9.9
 × 2.7

(3)   2.9
 × 6.6

(4)   5.3
 × 4.4

(5)   1.5
 × 8.9

(6)   7.9
 × 1.8

(7)   6.7
 × 5.2

(8)   8.5
 × 4.7

(9)   3.6
 × 9.5

(10)   9.2
 × 8.0

(11)   6.6
 × 9.7

(12)   9.5
 × 0.2

(13)   2.2
 × 6.7

(14)   6.6
 × 4.2

(15)   8.5
 × 2.8

(16)   4.3
 × 8.8

(17)   2.3
 × 6.7

(18)   9.9
 × 3.8

(19)   5.7
 × 5.0

(20)   5.7
 × 0.9

Multiplying Decimals

1. $\begin{array}{r} 5.3 \\ \times\ 0.3 \\ \hline \end{array}$
2. $\begin{array}{r} 0.4 \\ \times\ 4.0 \\ \hline \end{array}$
3. $\begin{array}{r} 0.2 \\ \times\ 5.1 \\ \hline \end{array}$
4. $\begin{array}{r} 4.7 \\ \times\ 0.8 \\ \hline \end{array}$

5. $\begin{array}{r} 9.8 \\ \times\ 5.0 \\ \hline \end{array}$
6. $\begin{array}{r} 3.7 \\ \times\ 6.4 \\ \hline \end{array}$
7. $\begin{array}{r} 6.5 \\ \times\ 1.9 \\ \hline \end{array}$
8. $\begin{array}{r} 6.2 \\ \times\ 6.9 \\ \hline \end{array}$

9. $\begin{array}{r} 0.4 \\ \times\ 5.7 \\ \hline \end{array}$
10. $\begin{array}{r} 2.5 \\ \times\ 1.9 \\ \hline \end{array}$
11. $\begin{array}{r} 1.2 \\ \times\ 4.3 \\ \hline \end{array}$
12. $\begin{array}{r} 0.9 \\ \times\ 8.8 \\ \hline \end{array}$

13. $\begin{array}{r} 5.3 \\ \times\ 4.3 \\ \hline \end{array}$
14. $\begin{array}{r} 7.5 \\ \times\ 1.0 \\ \hline \end{array}$
15. $\begin{array}{r} 4.3 \\ \times\ 3.8 \\ \hline \end{array}$
16. $\begin{array}{r} 3.2 \\ \times\ 8.6 \\ \hline \end{array}$

17. $\begin{array}{r} 2.2 \\ \times\ 2.5 \\ \hline \end{array}$
18. $\begin{array}{r} 2.9 \\ \times\ 7.8 \\ \hline \end{array}$
19. $\begin{array}{r} 6.0 \\ \times\ 0.8 \\ \hline \end{array}$
20. $\begin{array}{r} 0.9 \\ \times\ 2.9 \\ \hline \end{array}$

Multiplying Decimals

$\begin{array}{r} 1 \quad\ 2.3 \\ \times\ 5.8 \\ \hline \end{array}$ $\begin{array}{r} 2 \quad\ 9.9 \\ \times\ 1.1 \\ \hline \end{array}$ $\begin{array}{r} 3 \quad\ 7.1 \\ \times\ 4.7 \\ \hline \end{array}$ $\begin{array}{r} 4 \quad\ 7.7 \\ \times\ 9.5 \\ \hline \end{array}$

$\begin{array}{r} 5 \quad\ 0.5 \\ \times\ 4.5 \\ \hline \end{array}$ $\begin{array}{r} 6 \quad\ 0.8 \\ \times\ 3.4 \\ \hline \end{array}$ $\begin{array}{r} 7 \quad\ 7.2 \\ \times\ 5.2 \\ \hline \end{array}$ $\begin{array}{r} 8 \quad\ 8.4 \\ \times\ 2.0 \\ \hline \end{array}$

$\begin{array}{r} 9 \quad\ 10.0 \\ \times\ 2.3 \\ \hline \end{array}$ $\begin{array}{r} 10 \quad\ 4.3 \\ \times\ 5.1 \\ \hline \end{array}$ $\begin{array}{r} 11 \quad\ 3.3 \\ \times\ 7.9 \\ \hline \end{array}$ $\begin{array}{r} 12 \quad\ 10.0 \\ \times\ 1.3 \\ \hline \end{array}$

$\begin{array}{r} 13 \quad\ 0.9 \\ \times\ 3.7 \\ \hline \end{array}$ $\begin{array}{r} 14 \quad\ 4.3 \\ \times\ 3.5 \\ \hline \end{array}$ $\begin{array}{r} 15 \quad\ 0.5 \\ \times\ 8.5 \\ \hline \end{array}$ $\begin{array}{r} 16 \quad\ 8.6 \\ \times\ 8.0 \\ \hline \end{array}$

$\begin{array}{r} 17 \quad\ 5.8 \\ \times\ 2.0 \\ \hline \end{array}$ $\begin{array}{r} 18 \quad\ 4.4 \\ \times\ 5.9 \\ \hline \end{array}$ $\begin{array}{r} 19 \quad\ 6.7 \\ \times\ 1.2 \\ \hline \end{array}$ $\begin{array}{r} 20 \quad\ 9.7 \\ \times\ 8.7 \\ \hline \end{array}$

Multiplying Decimals

①   4.3
   ×   5.5
___

②   0.7
   ×   2.1
___

③   9.9
   ×   5.9
___

④   4.2
   ×   1.1
___

⑤   0.6
   ×   0.2
___

⑥   0.5
   ×   3.5
___

⑦   7.8
   ×   8.0
___

⑧   4.2
   ×   6.9
___

⑨   5.5
   ×   7.8
___

⑩   8.3
   ×   9.5
___

⑪   0.4
   ×   6.8
___

⑫   7.0
   ×   1.1
___

⑬   3.7
   ×   2.9
___

⑭   1.8
   ×   8.6
___

⑮   2.3
   ×   6.3
___

⑯   2.8
   ×   2.1
___

⑰   9.3
   ×   9.9
___

⑱   3.5
   ×   5.4
___

⑲   9.5
   ×   8.3
___

⑳   8.0
   ×   5.8
___

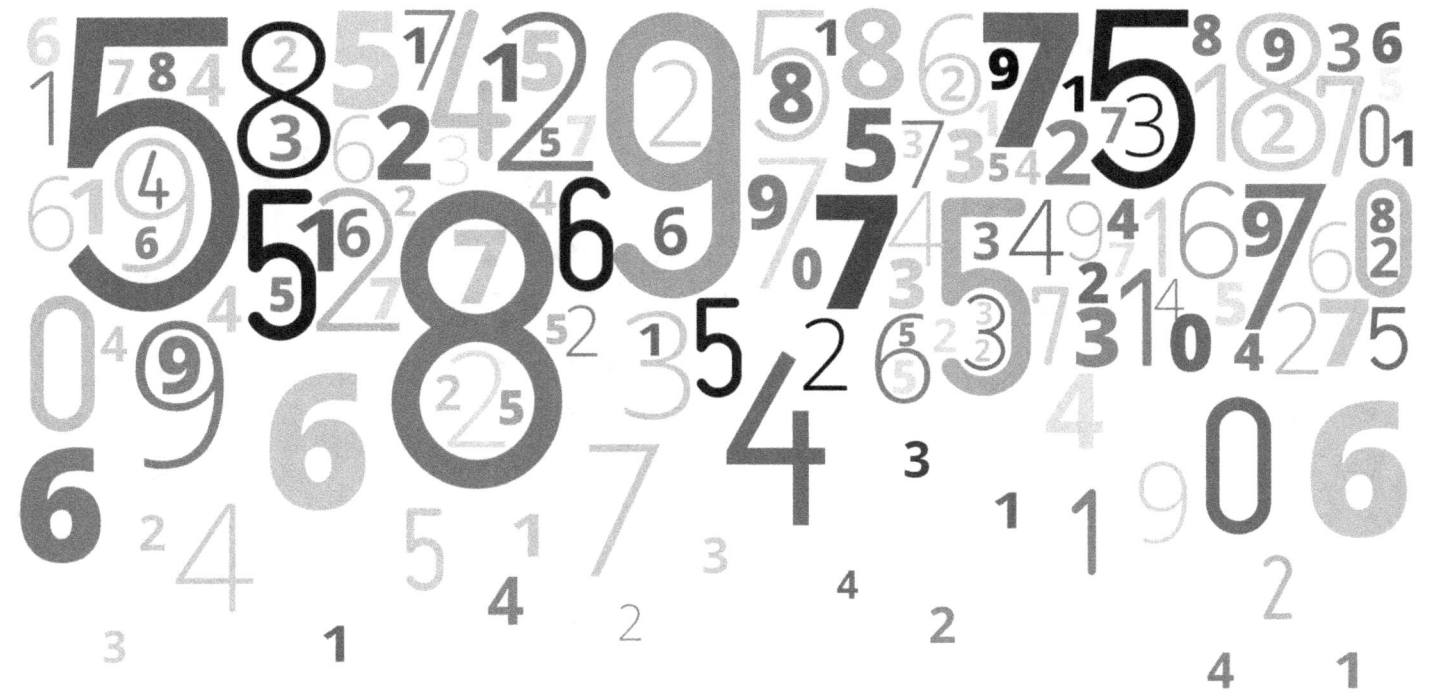

Solutions

Page 6, Item 1:

(1)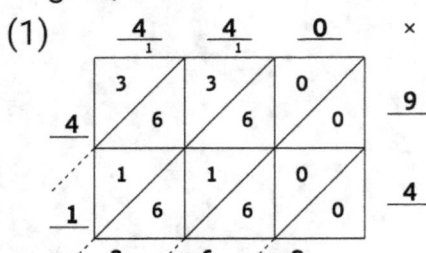

440 × 94 = 41,360

(2)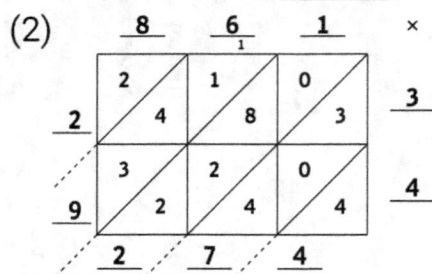

861 × 34 = 29,274

(3)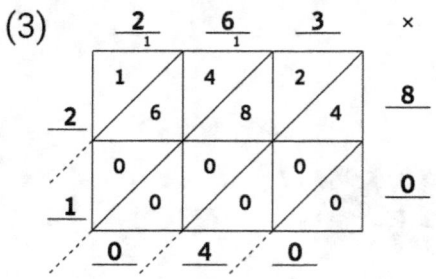

263 × 80 = 21,040

(4)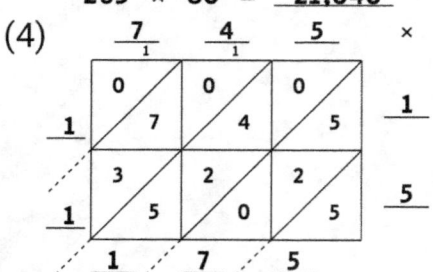

745 × 15 = 11,175

(5)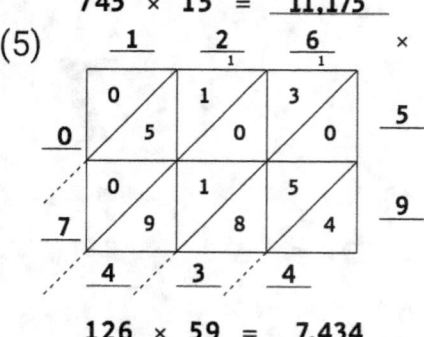

126 × 59 = 7,434

(6)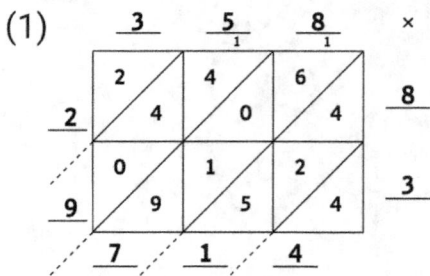

868 × 20 = 17,360

Page 7, Item 1:

(1)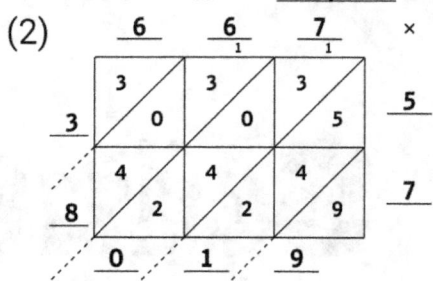

358 × 83 = 29,714

(2)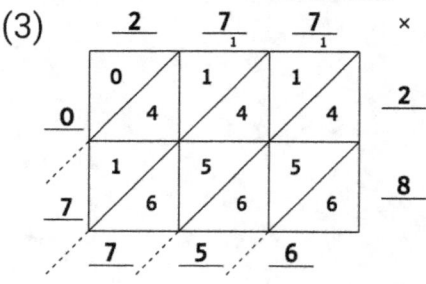

667 × 57 = 38,019

(3)

277 × 28 = 7,756

(4)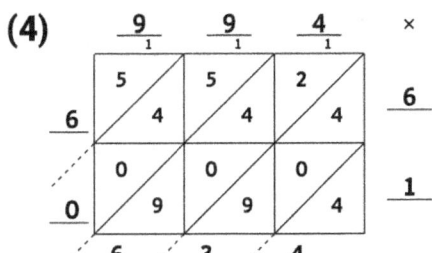

994 × 61 = __60,634__

(5)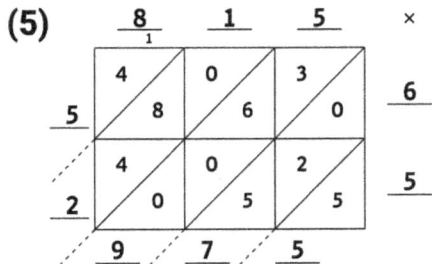

815 × 65 = __52,975__

(6)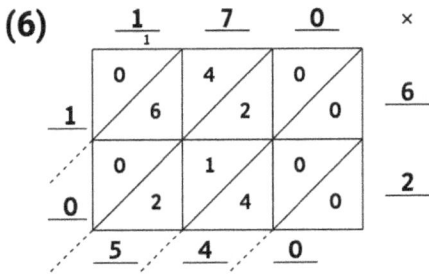

170 × 62 = __10,540__

Page 8, Item 1:

(1)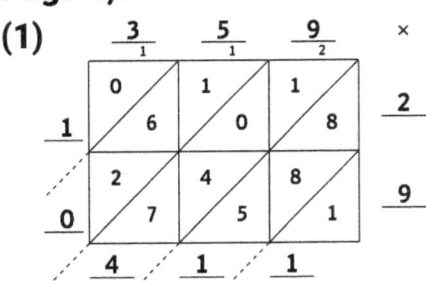

359 × 29 = __10,411__

(2)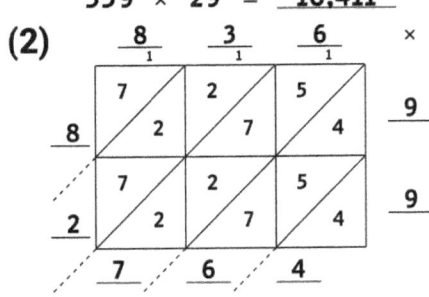

836 × 99 = __82,764__

(3)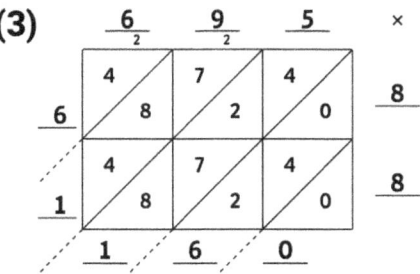

695 × 88 = __61,160__

(4)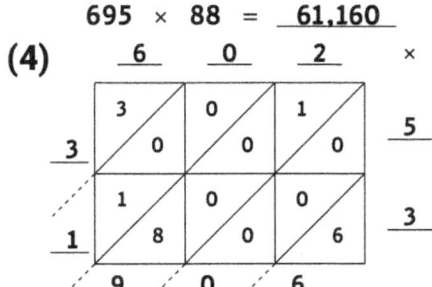

602 × 53 = __31,906__

(5)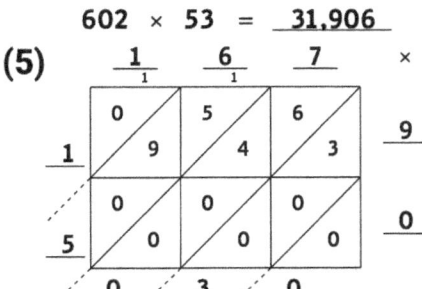

167 × 90 = __15,030__

(6)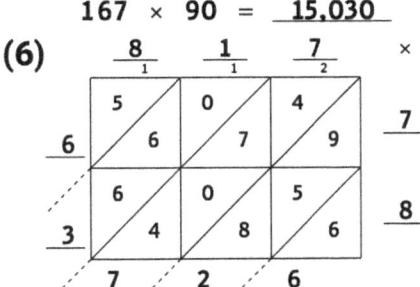

817 × 78 = __63,726__

Page 9, Item 1:

(1)

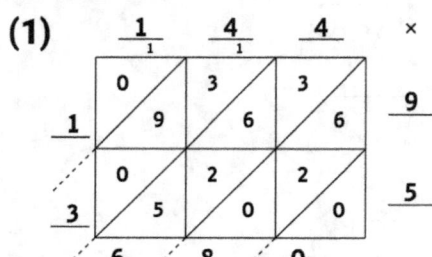

144 × 95 = __13,680__

(2)

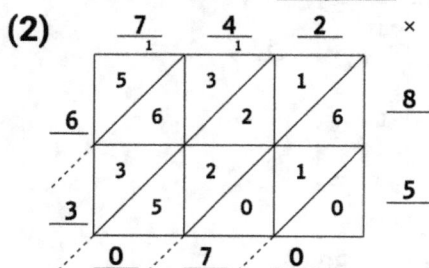

742 × 85 = __63,070__

(3)

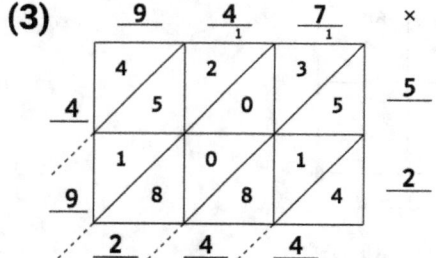

947 × 52 = __49,244__

(4)

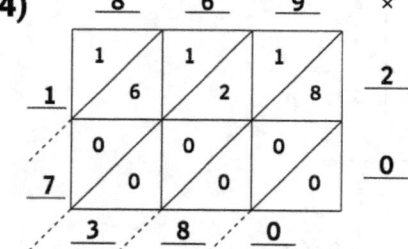

869 × 20 = __17,380__

(5)

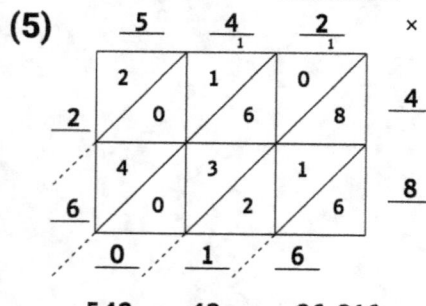

542 × 48 = __26,016__

(6)

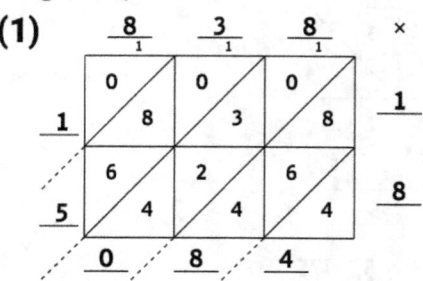

781 × 51 = __39,831__

Page 10, Item 1:

(1)

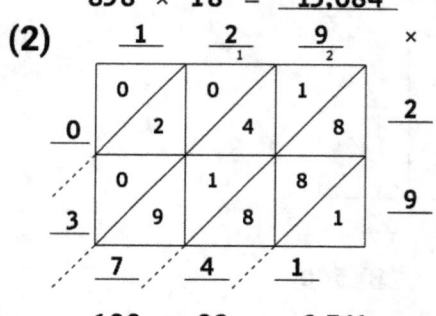

838 × 18 = __15,084__

(2)

129 × 29 = __3,741__

108

(3)

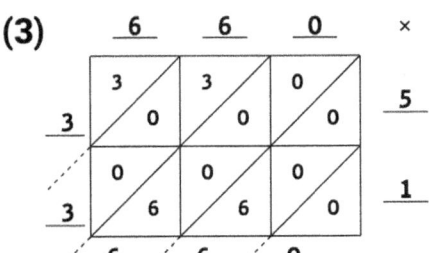

660 × 51 = **33,660**

(4)

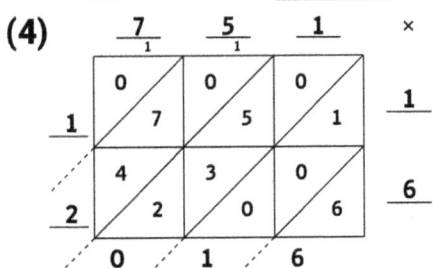

751 × 16 = **12,016**

(5)

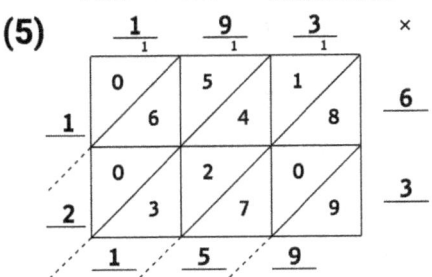

193 × 63 = **12,159**

(6)

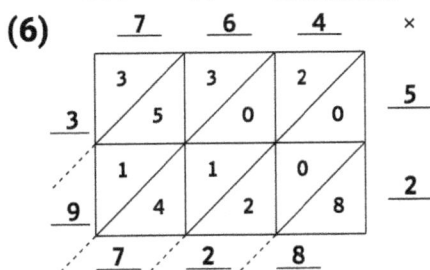

764 × 52 = **39,728**

Page 11, Item 1:

(1)

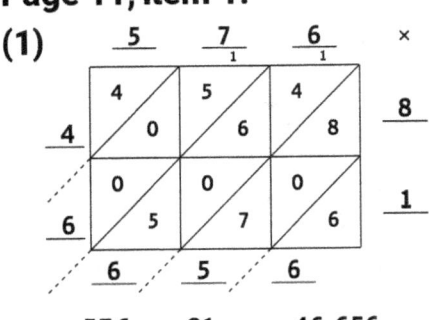

576 × 81 = **46,656**

(2)

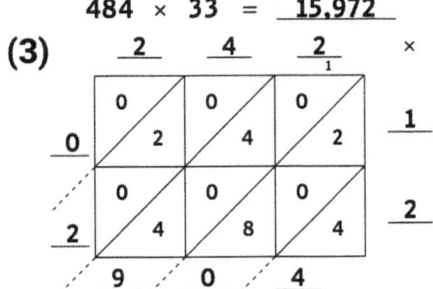

484 × 33 = **15,972**

(3)

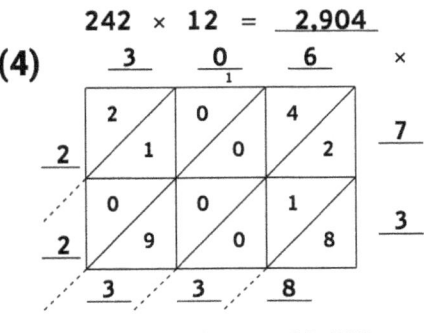

242 × 12 = **2,904**

(4)

306 × 73 = **22,338**

(5)

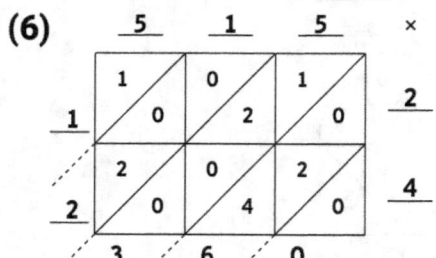

$211 \times 22 = \underline{4,642}$

(6)

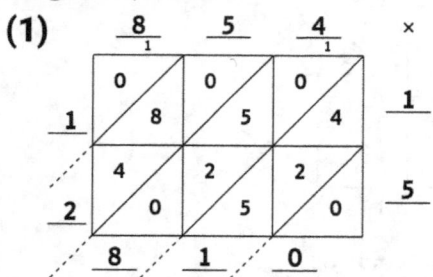

$515 \times 24 = \underline{12,360}$

Page 12, Item 1:

(1)

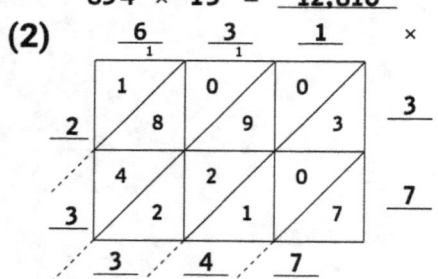

$854 \times 15 = \underline{12,810}$

(2)

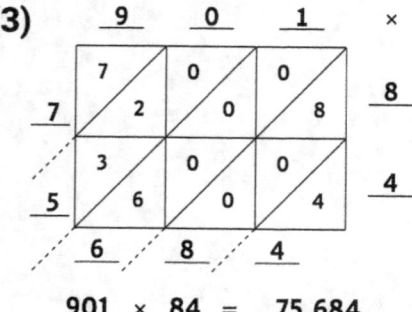

$631 \times 37 = \underline{23,347}$

(3)

$901 \times 84 = \underline{75,684}$

(4)

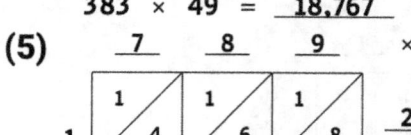
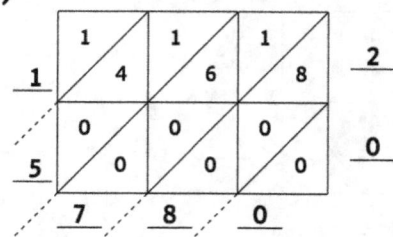

$383 \times 49 = \underline{18,767}$

(5)

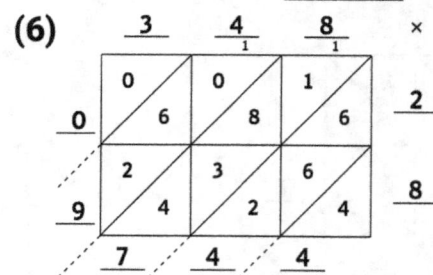

$789 \times 20 = \underline{15,780}$

(6)

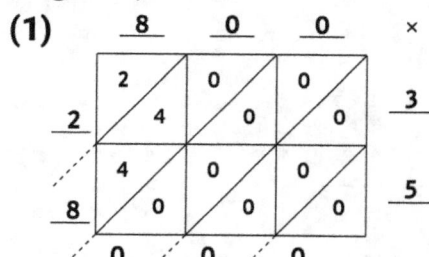

$348 \times 28 = \underline{9,744}$

Page 13, Item 1:

(1)

$800 \times 35 = \underline{28,000}$

(2)

$808 \times 54 = \underline{43,632}$

(3)

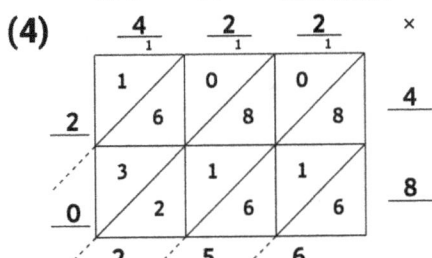

633 × 15 = __9,495__

(4)

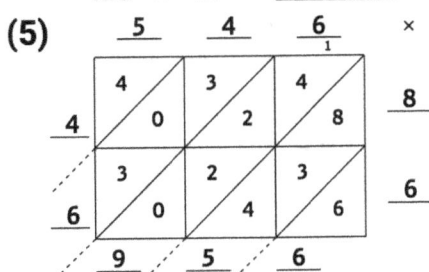

422 × 48 = __20,256__

(5)

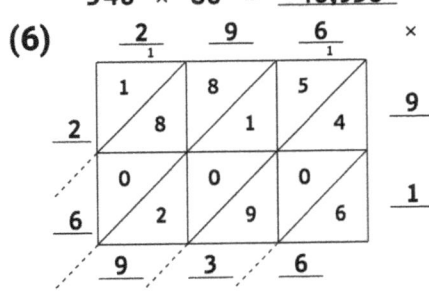

546 × 86 = __46,956__

(6)

296 × 91 = __26,936__

Page 14, Item 1:

(1)

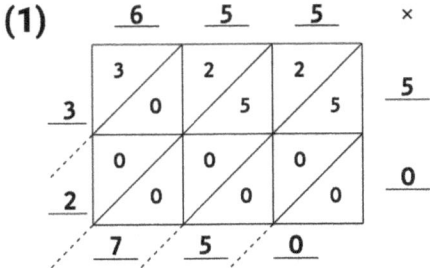

655 × 50 = __32,750__

(2)

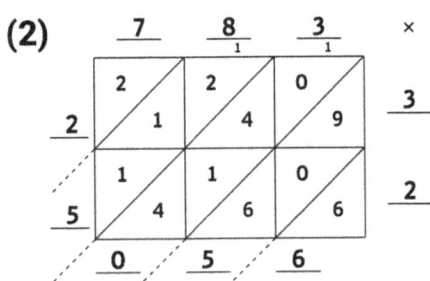

783 × 32 = __25,056__

(3)

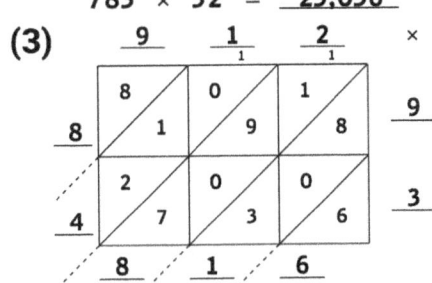

912 × 93 = __84,816__

(4)

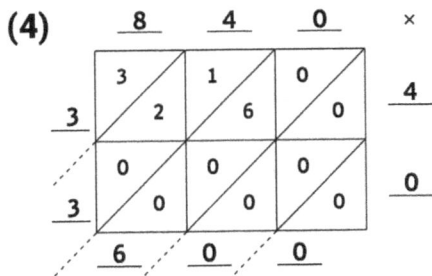

840 × 40 = __33,600__

111

(5)

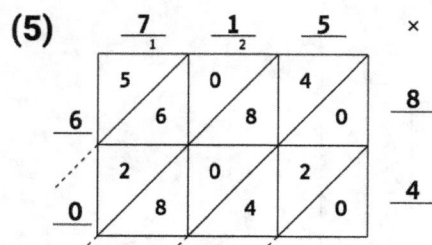

715 × 84 = <u>60,060</u>

(6)

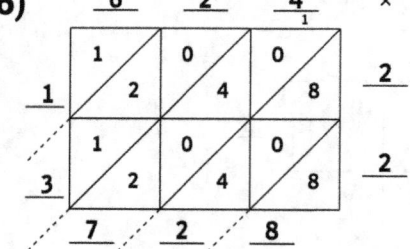

624 × 22 = <u>13,728</u>

Page 15, Item 1:

(1)

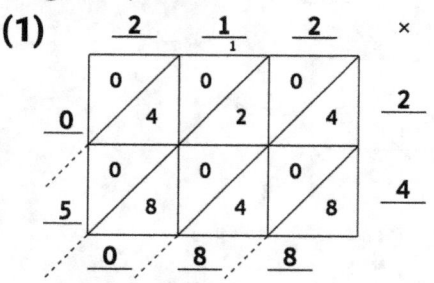

212 × 24 = <u>5,088</u>

(2)

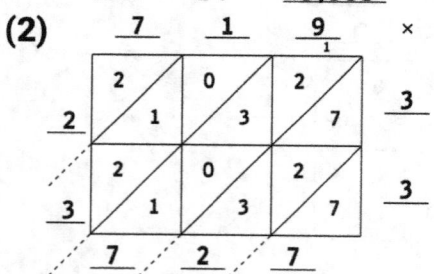

719 × 33 = <u>23,727</u>

(3)

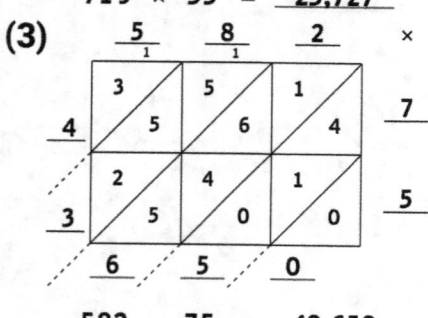

582 × 75 = <u>43,650</u>

(4)

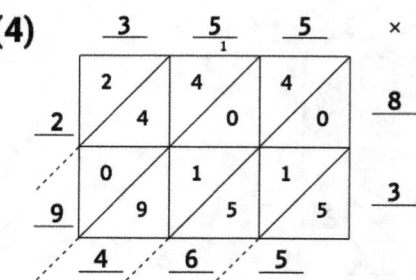

355 × 83 = <u>29,465</u>

(5)

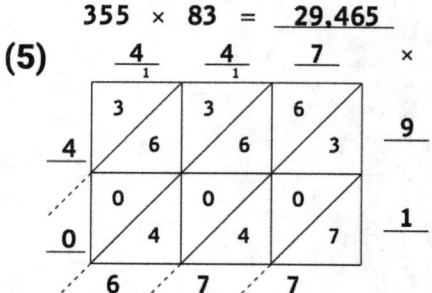

447 × 91 = <u>40,677</u>

(6)

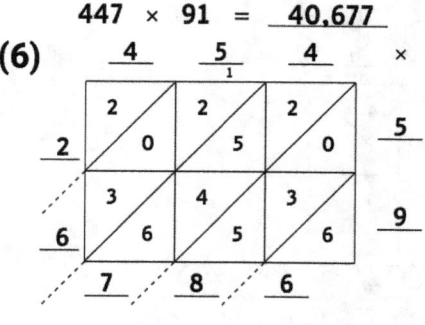

454 × 59 = <u>26,786</u>

112

Page 16, Item 1:

(1)

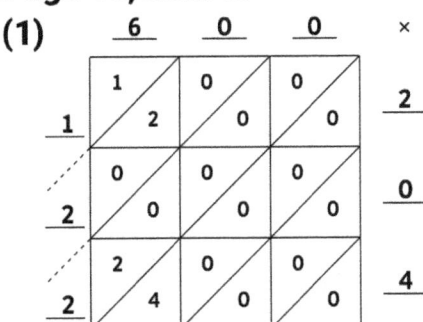

$600 \times 204 = \underline{122,400}$

(2)

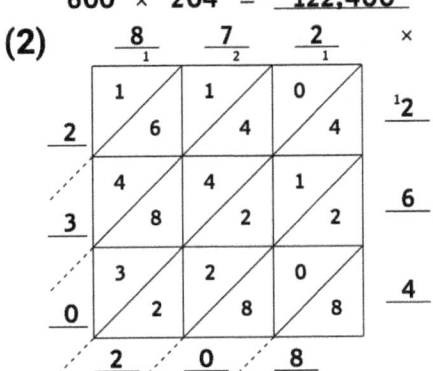

$872 \times 264 = \underline{230,208}$

(3)

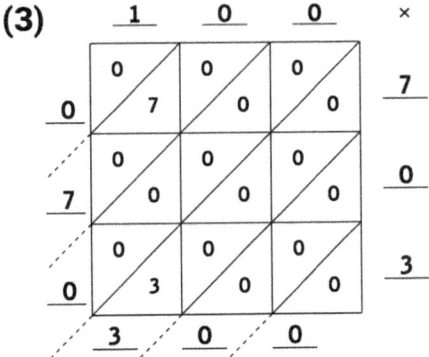

$100 \times 703 = \underline{70,300}$

(4)

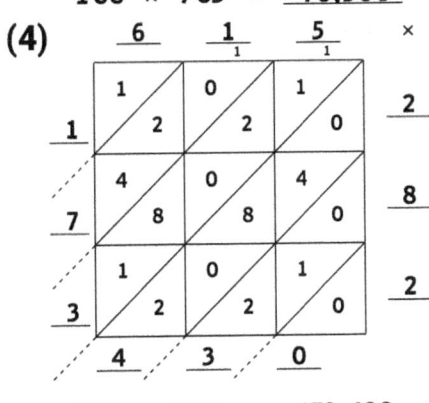

$615 \times 282 = \underline{173,430}$

Page 17, Item 1:

(1)

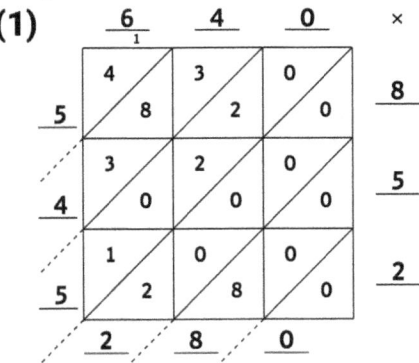

$640 \times 852 = \underline{545,280}$

(2)

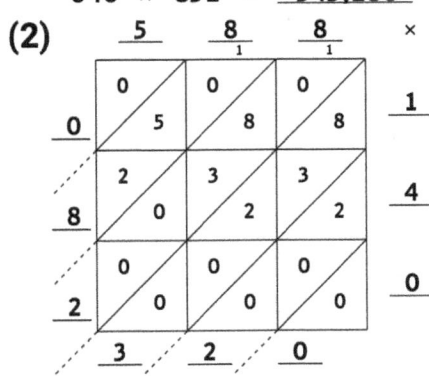

$588 \times 140 = \underline{82,320}$

(3)

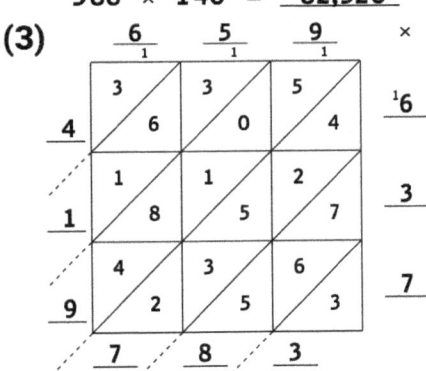

$659 \times 637 = \underline{419,783}$

(4)

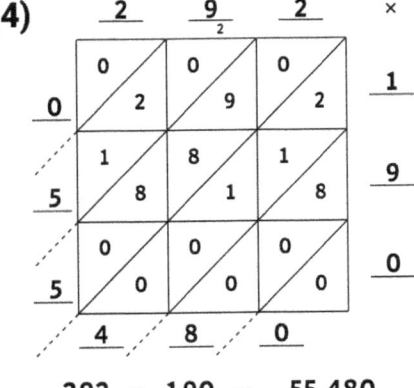

$292 \times 190 = \underline{55,480}$

Page 18, Item 1:

(1)
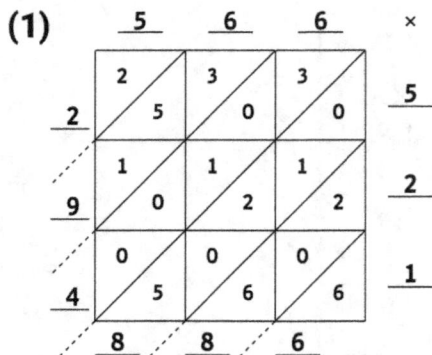

566 × 521 = __294,886__

(2)

599 × 925 = __554,075__

(3)

472 × 164 = __77,408__

(4)

970 × 882 = __855,540__

Page 19, Item 1:

(1)
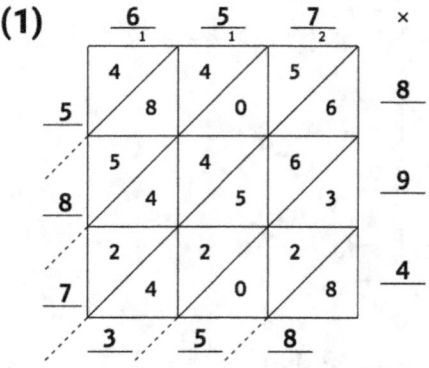

657 × 894 = __587,358__

(2)
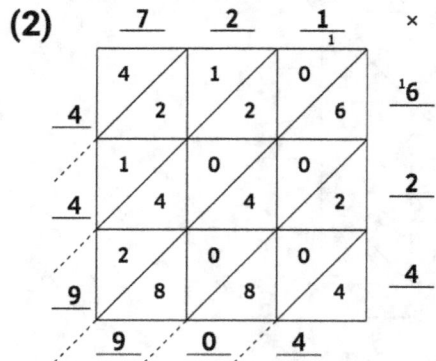

721 × 624 = __449,904__

(3)
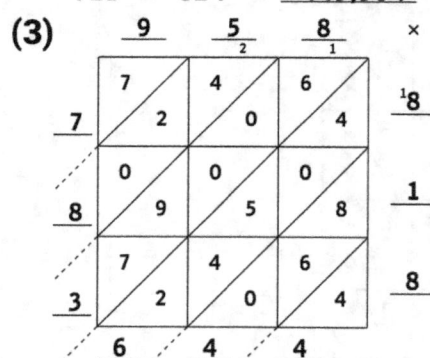

958 × 818 = __783,644__

(4)
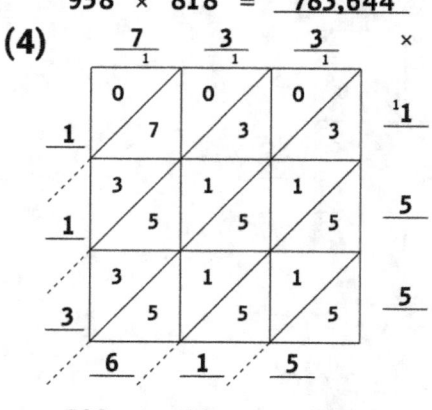

733 × 155 = __113,615__

Page 20, Item 1:

(1)

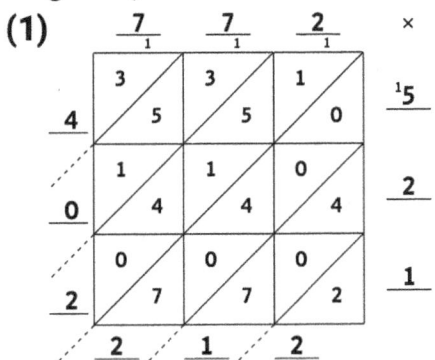

772 × 521 = __402,212__

(2)

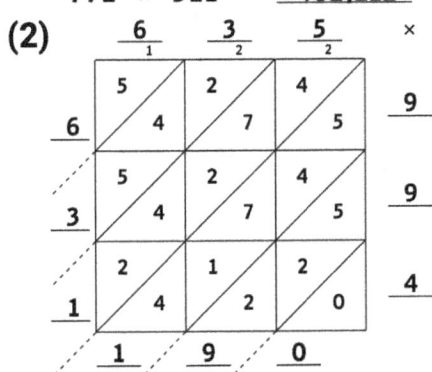

635 × 994 = __631,190__

(3)

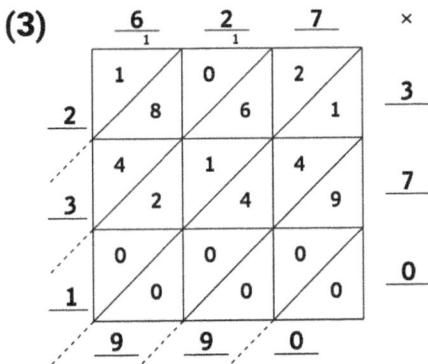

627 × 370 = __231,990__

(4)

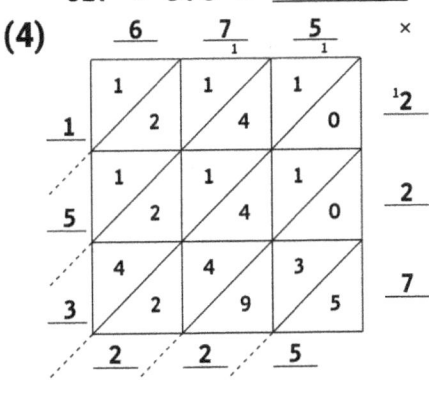

675 × 227 = __153,225__

Page 21, Item 1:

(1)

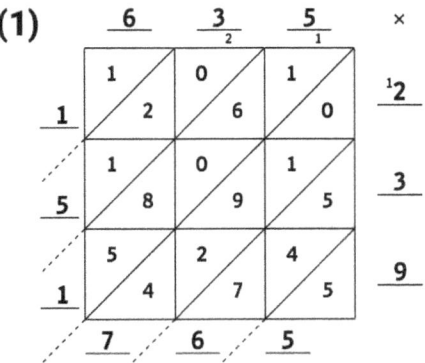

635 × 239 = __151,765__

(2)

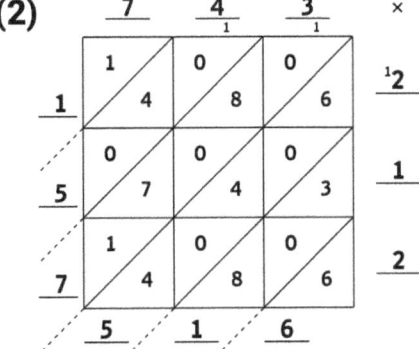

743 × 212 = __157,516__

(3)

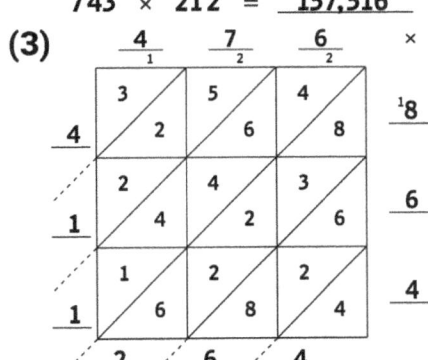

476 × 864 = __411,264__

(4)

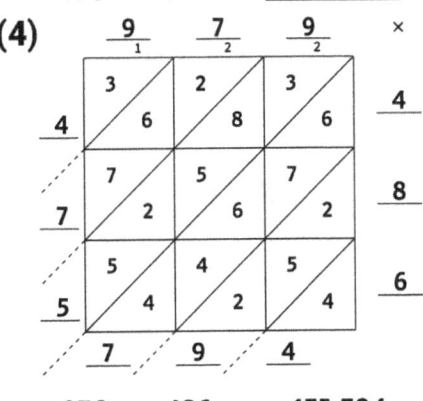

979 × 486 = __475,794__

Page 22, Item 1:

(1)

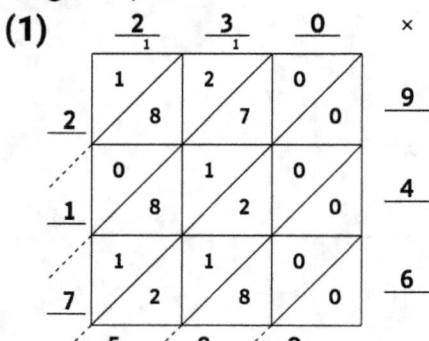

$230 \times 946 = \underline{217,580}$

(2)

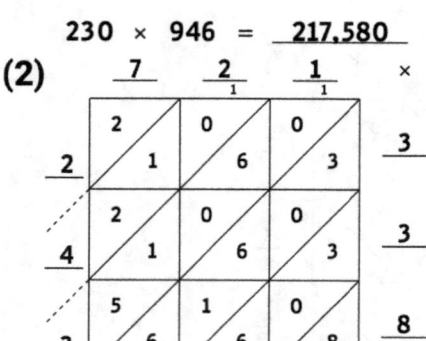

$721 \times 338 = \underline{243,698}$

(3)

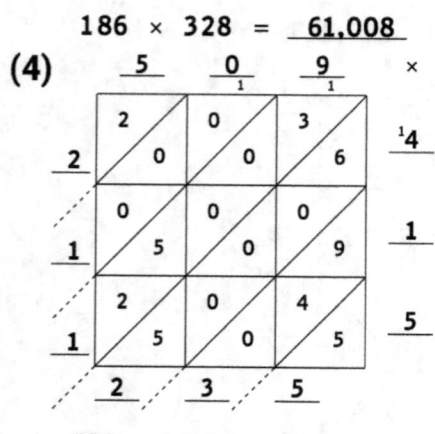

$186 \times 328 = \underline{61,008}$

(4)

$509 \times 415 = \underline{211,235}$

Page 23, Item 1:

(1)

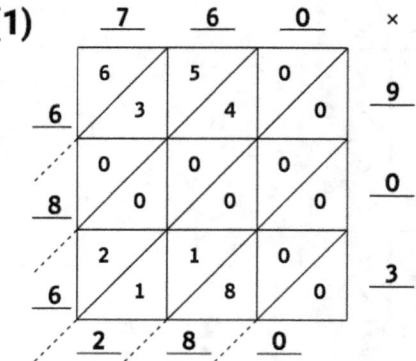

$760 \times 903 = \underline{686,280}$

(2)

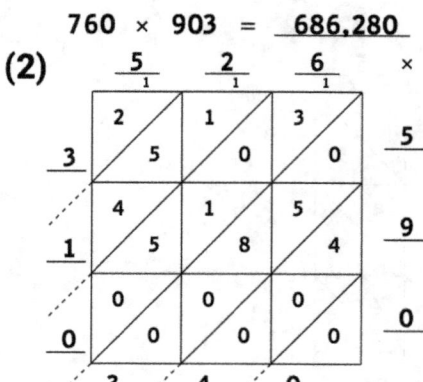

$526 \times 590 = \underline{310,340}$

(3)

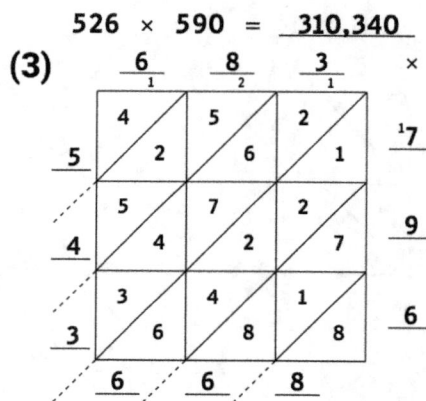

$683 \times 796 = \underline{543,668}$

(4)

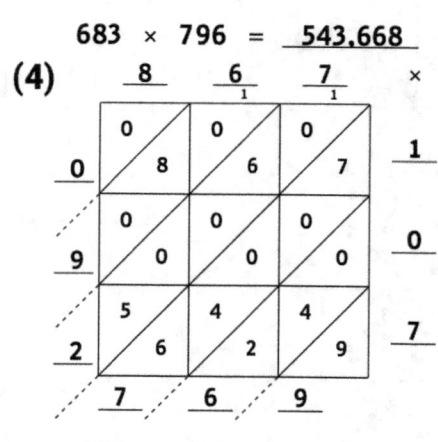

$867 \times 107 = \underline{92,769}$

Page 24, Item 1:

(1)

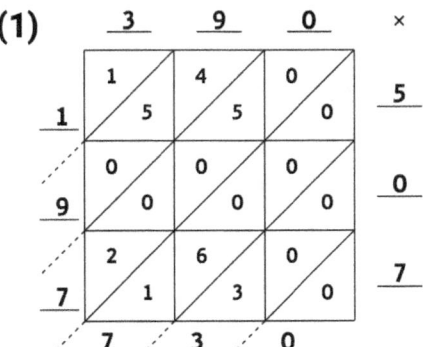

$$390 \times 507 = \underline{197{,}730}$$

(2)

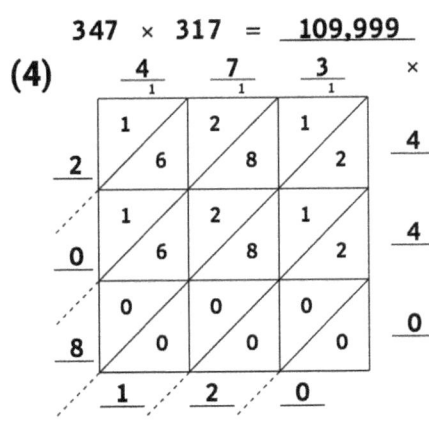

$$925 \times 467 = \underline{431{,}975}$$

(3)

$$347 \times 317 = \underline{109{,}999}$$

(4)

$$473 \times 440 = \underline{208{,}120}$$

Page 25, Item 1:

(1)

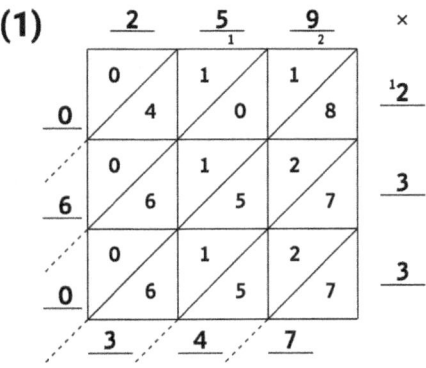

$$259 \times 233 = \underline{60{,}347}$$

(2)

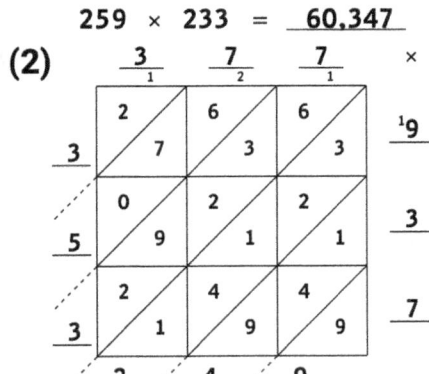

$$377 \times 937 = \underline{353{,}249}$$

(3)

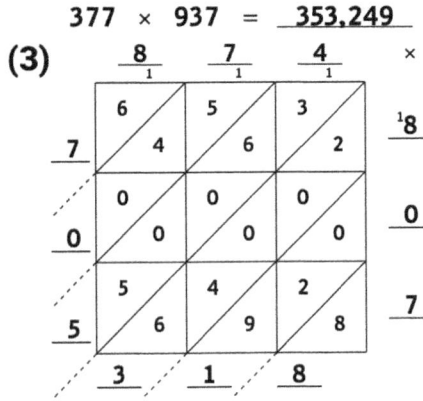

$$874 \times 807 = \underline{705{,}318}$$

(4)

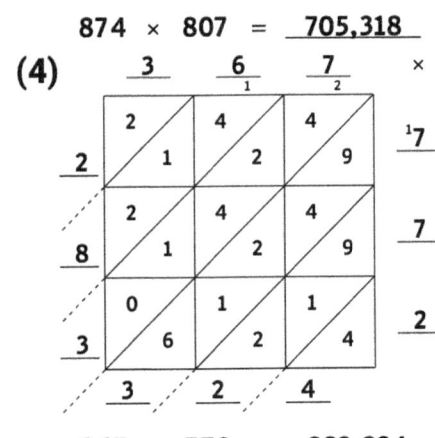

$$367 \times 772 = \underline{283{,}324}$$

Page 26, Item 1:

(1)

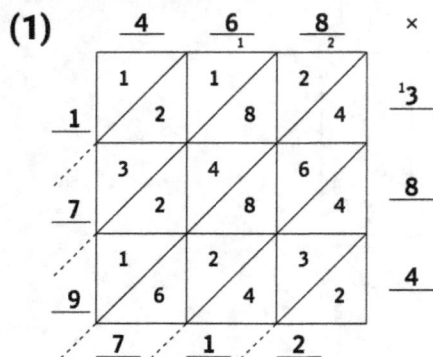

$468 \times 384 = \underline{179{,}712}$

(2)

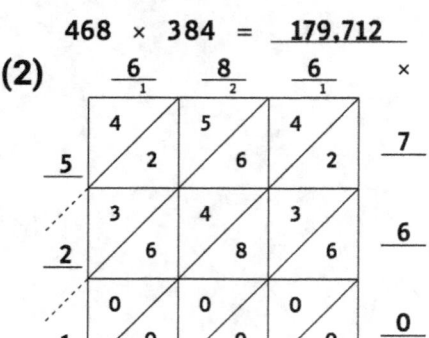

$686 \times 760 = \underline{521{,}360}$

(3)

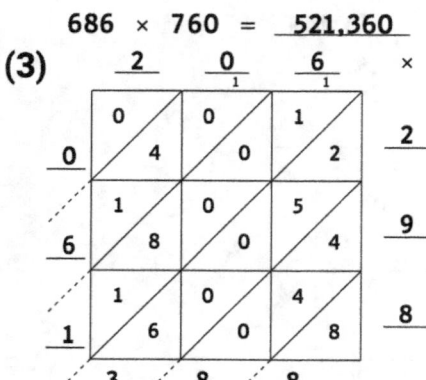

$206 \times 298 = \underline{61{,}388}$

(4)

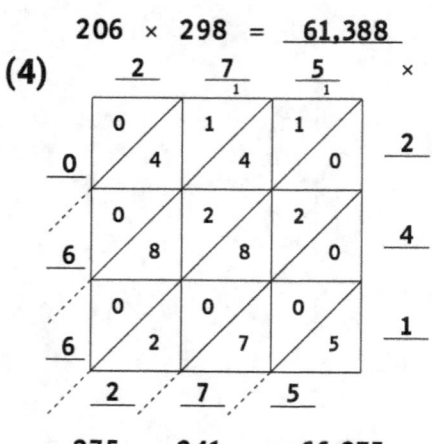

$275 \times 241 = \underline{66{,}275}$

Page 27, Item 1:

(1)

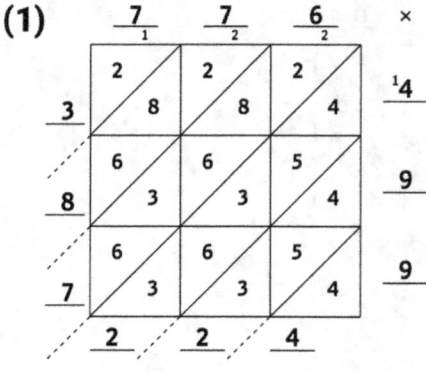

$776 \times 499 = \underline{387{,}224}$

(2)

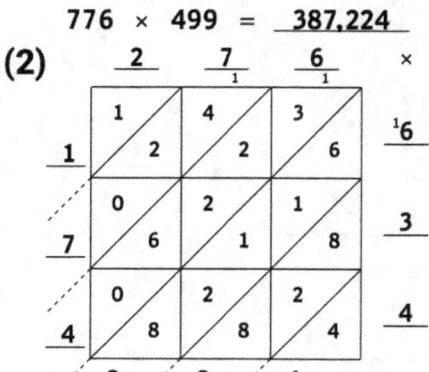

$276 \times 634 = \underline{174{,}984}$

(3)

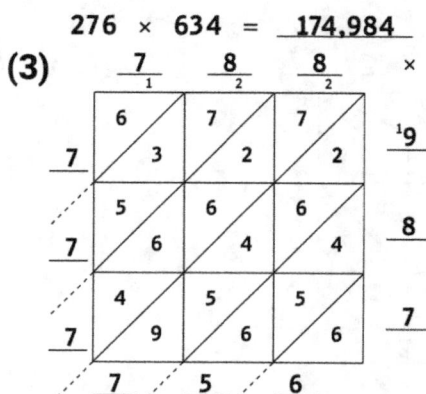

$788 \times 987 = \underline{777{,}756}$

(4)

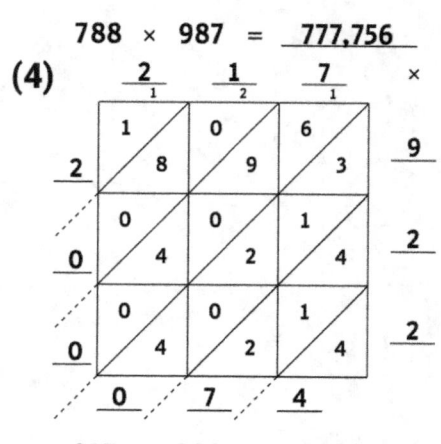

$217 \times 922 = \underline{200{,}074}$

118

Page 28, Item 1:

(1)

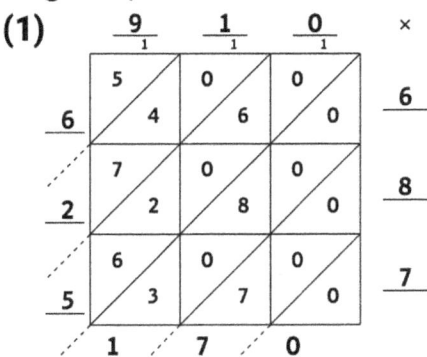

910 × 687 = 625,170

(2)

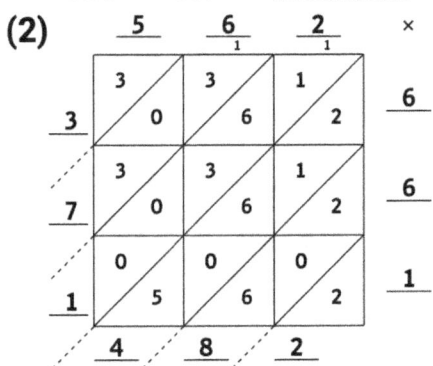

562 × 661 = 371,482

(3)

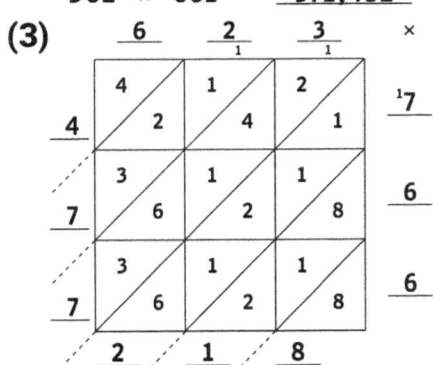

623 × 766 = 477,218

(4)

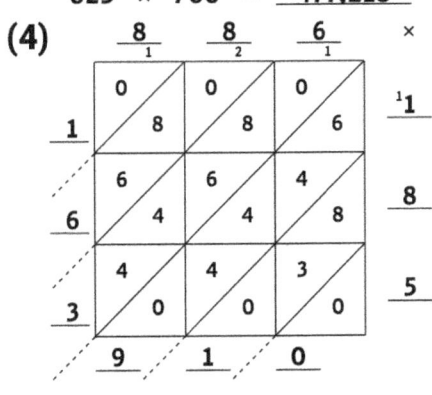

886 × 185 = 163,910

Page 29, Item 1:

(1)

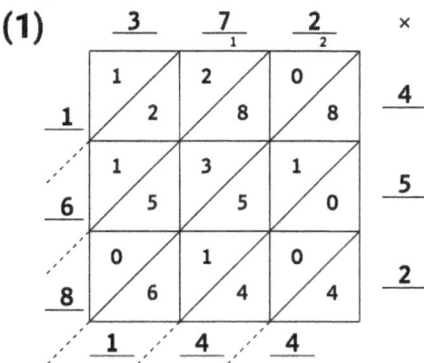

372 × 452 = 168,144

(2)

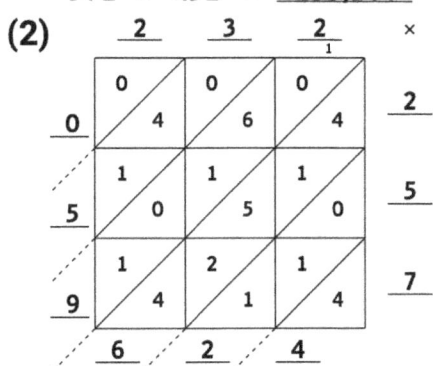

232 × 257 = 59,624

(3)

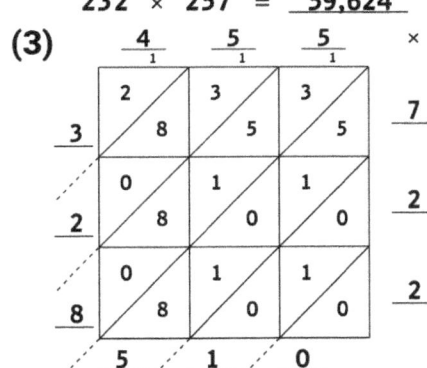

455 × 722 = 328,510

(4)

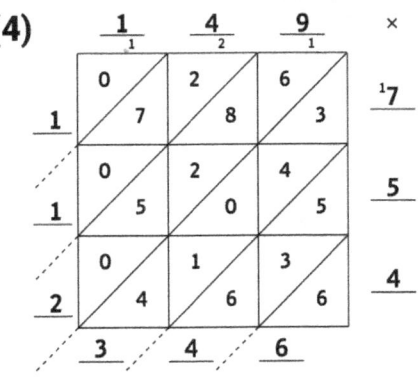

149 × 754 = 112,346

Page 30, Item 1:

(1)

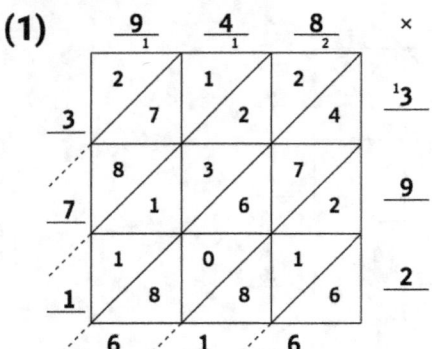

$948 \times 392 = \underline{371,616}$

(2)

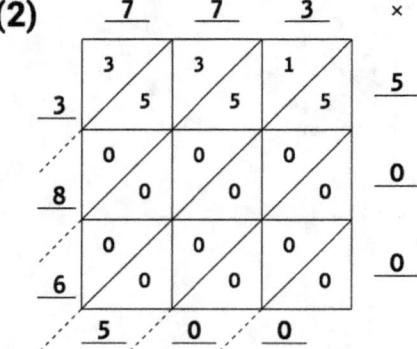

$773 \times 500 = \underline{386,500}$

(3)

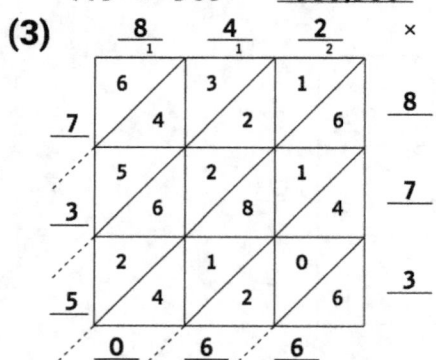

$842 \times 873 = \underline{735,066}$

(4)

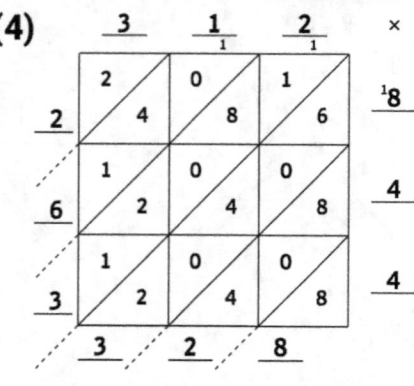

$312 \times 844 = \underline{263,328}$

Page 32, Item 1:

(1)
```
        3 0 R3
  8 ) 2 4 3
    - 2 4
        0 3
      -   0
          3
```

(2)
```
        2 7 2 R2
  3 ) 8 1 8
    - 6
      2 1
    - 2 1
        0 8
      -   6
          2
```

(3)
```
        1 0 7 R1
  3 ) 3 2 2
    - 3
      0 2
    -   0
        2 2
      - 2 1
          1
```

(4)
```
        2 8 6 R1
  3 ) 8 5 9
    - 6
      2 5
    - 2 4
        1 9
      - 1 8
          1
```

(5)
```
        6 1 R2
  3 ) 1 8 5
    - 1 8
        0 5
      -   3
          2
```

(6)
```
        3 1 R1
  5 ) 1 5 6
    - 1 5
        0 6
      -   5
          1
```

(7)
```
        1 2 4 R3
  6 ) 7 4 7
    - 6
      1 4
    - 1 2
        2 7
      - 2 4
          3
```

(8)
```
        1 0 2 R0
  5 ) 5 1 0
    - 5
      0 1
    -   0
        1 0
      - 1 0
          0
```

(9)
```
        3 6 0 R0
  2 ) 7 2 0
    - 6
      1 2
    - 1 2
        0 0
      -   0
          0
```

120

Page 33, Item 1:

(1)
```
      1 6 8 R4
  5 ) 8 4 4
    - 5
      3 4
    - 3 0
        4 4
      - 4 0
          4
```

(2)
```
        9 9 R2
  4 ) 3 9 8
    - 3 6
        3 8
      - 3 6
          2
```

(3)
```
        4 4 R5
  7 ) 3 1 3
    - 2 8
        3 3
      - 2 8
          5
```

(4)
```
      1 7 8 R4
  5 ) 8 9 4
    - 5
      3 9
    - 3 5
        4 4
      - 4 0
          4
```

(5)
```
      2 3 6 R1
  2 ) 4 7 3
    - 4
      0 7
    -   6
        1 3
      - 1 2
          1
```

(6)
```
      1 2 6 R5
  7 ) 8 8 7
    - 7
      1 8
    - 1 4
        4 7
      - 4 2
          5
```

(7)
```
        7 3 R7
  9 ) 6 6 4
    - 6 3
        3 4
      - 2 7
          7
```

(8)
```
      1 1 2 R3
  7 ) 7 8 7
    - 7
      0 8
    -   7
        1 7
      - 1 4
          3
```

(9)
```
        5 5 R3
  4 ) 2 2 3
    - 2 0
        2 3
      - 2 0
          3
```

Page 34, Item 1:

(1)
```
      1 1 9 R2
  6 ) 7 1 6
    - 6
      1 1
    -   6
        5 6
      - 5 4
          2
```

(2)
```
        9 5 R2
  5 ) 4 7 7
    - 4 5
        2 7
      - 2 5
          2
```

(3)
```
      4 3 7 R1
  2 ) 8 7 5
    - 8
      0 7
    -   6
        1 5
      - 1 4
          1
```

(4)
```
        2 9 R1
  9 ) 2 6 2
    - 1 8
        8 2
      - 8 1
          1
```

(5)
```
        9 7 R0
  9 ) 8 7 3
    - 8 1
        6 3
      - 6 3
          0
```

(6)
```
      3 1 4 R0
  3 ) 9 4 2
    - 9
      0 4
    -   3
        1 2
      - 1 2
          0
```

(7)
```
      1 3 2 R3
  6 ) 7 9 5
    - 6
      1 9
    - 1 8
        1 5
      - 1 2
          3
```

(8)
```
        1 6 R0
  8 ) 1 2 8
    - 8
      4 8
    - 4 8
        0
```

(9)
```
        7 7 R0
  9 ) 6 9 3
    - 6 3
        6 3
      - 6 3
          0
```

Page 35, Item 1:

(1)
```
      2 3 7 R0
  4 ) 9 4 8
    - 8
      1 4
    - 1 2
        2 8
      - 2 8
          0
```

(2)
```
      1 0 9 R4
  5 ) 5 4 9
    - 5
      0 4
    -   0
        4 9
      - 4 5
          4
```

(3)
```
        8 0 R2
  7 ) 5 6 2
    - 5 6
      0 2
    -   0
        2
```

(4)
```
        7 7 R1
  9 ) 6 9 4
    - 6 3
        6 4
      - 6 3
          1
```

(5)
```
        3 8 R7
  9 ) 3 4 9
    - 2 7
        7 9
      - 7 2
          7
```

(6)
```
      3 7 2 R0
  2 ) 7 4 4
    - 6
      1 4
    - 1 4
        0 4
      -   4
          0
```

(7)
```
      3 3 4 R0
  2 ) 6 6 8
    - 6
      0 6
    -   6
        0 8
      -   8
          0
```

(8)
```
      3 9 3 R1
  2 ) 7 8 7
    - 6
      1 8
    - 1 8
        0 7
      -   6
          1
```

(9)
```
      1 9 7 R1
  3 ) 5 9 2
    - 3
      2 9
    - 2 7
        2 2
      - 2 1
          1
```

Page 36, Item 1:

(1)
```
      2 0 5 R3
  4 ) 8 2 3
    - 8
      0 2
    -   0
        2 3
      - 2 0
          3
```

(2)
```
        5 8 R5
  8 ) 4 6 9
    - 4 0
        6 9
      - 6 4
          5
```

(3)
```
        6 6 R3
  6 ) 3 9 9
    - 3 6
        3 9
      - 3 6
          3
```

(4)
```
      1 9 4 R0
  3 ) 5 8 2
    - 3
      2 8
    - 2 7
        1 2
      - 1 2
          0
```

(5)
```
      1 9 6 R0
  5 ) 9 8 0
    - 5
      4 8
    - 4 5
        3 0
      - 3 0
          0
```

(6)
```
      1 2 4 R1
  7 ) 8 6 9
    - 7
      1 6
    - 1 4
        2 9
      - 2 8
          1
```

(7)
```
        2 5 R1
  6 ) 1 5 1
    - 1 2
        3 1
      - 3 0
          1
```

(8)
```
      1 0 4 R3
  8 ) 8 3 5
    - 8
      0 3
    -   0
        3 5
      - 3 2
          3
```

(9)
```
        5 4 R0
  4 ) 2 1 6
    - 2 0
        1 6
      - 1 6
          0
```

Page 37, Item 1:

(1)
```
    3 0 4 R2
3 ) 9 1 4
  - 9
  -----
    0 1
  -   0
  -----
      1 4
  -   1 2
  -------
        2
```

(2)
```
    1 9 8 R3
4 ) 7 9 5
  - 4
  -----
    3 9
  - 3 6
  -----
      3 5
  -   3 2
  -------
        3
```

(3)
```
    4 5 4 R0
2 ) 9 0 8
  - 8
  -----
    1 0
  - 1 0
  -----
      0 8
  -     8
  -------
        0
```

(4)
```
    1 7 0 R0
3 ) 5 1 0
  - 3
  -----
    2 1
  - 2 1
  -----
      0 0
  -     0
  -------
        0
```

(5)
```
    3 7 5 R0
2 ) 7 5 0
  - 6
  -----
    1 5
  - 1 4
  -----
      1 0
  -   1 0
  -------
        0
```

(6)
```
      6 2 R0
7 ) 4 3 4
  - 4 2
  -----
      1 4
  -   1 4
  -------
        0
```

(7)
```
      9 2 R8
9 ) 8 3 6
  - 8 1
  -----
      2 6
  -   1 8
  -------
        8
```

(8)
```
      4 9 R2
7 ) 3 4 5
  - 2 8
  -----
      6 5
  -   6 3
  -------
        2
```

(9)
```
      2 6 R3
6 ) 1 5 9
  - 1 2
  -----
      3 9
  -   3 6
  -------
        3
```

Page 38, Item 1:

(1)
```
    1 1 1 R1
3 ) 3 3 4
  - 3
  -----
    0 3
  -   3
  -----
      0 4
  -     3
  -------
        1
```

(2)
```
    3 4 6 R1
2 ) 6 9 3
  - 6
  -----
    0 9
  -   8
  -----
      1 3
  -   1 2
  -------
        1
```

(3)
```
    1 1 4 R4
8 ) 9 1 6
  - 8
  -----
    1 1
  -   8
  -----
      3 6
  -   3 2
  -------
        4
```

(4)
```
      8 1 R0
7 ) 5 6 7
  - 5 6
  -----
      0 7
  -     7
  -------
        0
```

(5)
```
      8 6 R3
6 ) 5 1 9
  - 4 8
  -----
      3 9
  -   3 6
  -------
        3
```

(6)
```
    4 1 0 R1
2 ) 8 2 1
  - 8
  -----
    0 2
  -   2
  -----
      0 1
  -     0
  -------
        1
```

(7)
```
    1 7 6 R2
3 ) 5 3 0
  - 3
  -----
    2 3
  - 2 1
  -----
      2 0
  -   1 8
  -------
        2
```

(8)
```
      4 5 R2
9 ) 4 0 7
  - 3 6
  -----
      4 7
  -   4 5
  -------
        2
```

(9)
```
      3 9 R1
5 ) 1 9 6
  - 1 5
  -----
      4 6
  -   4 5
  -------
        1
```

123

Page 39, Item 1:

(1)
```
      1 6 8 R0
  3 ) 5 0 4
    - 3
      2 0
    - 1 8
        2 4
      - 2 4
          0
```

(2)
```
      1 7 3 R2
  3 ) 5 2 1
    - 3
      2 2
    - 2 1
        1 1
      -   9
          2
```

(3)
```
        9 7 R1
  9 ) 8 7 4
    - 8 1
        6 4
      - 6 3
          1
```

(4)
```
      1 4 1 R1
  2 ) 2 8 3
    - 2
      0 8
      - 8
        0 3
      -   2
          1
```

(5)
```
        8 5 R2
  9 ) 7 6 7
    - 7 2
        4 7
      - 4 5
          2
```

(6)
```
        8 7 R3
  5 ) 4 3 8
    - 4 0
        3 8
      - 3 5
          3
```

(7)
```
        7 2 R2
  6 ) 4 3 4
    - 4 2
        1 4
      - 1 2
          2
```

(8)
```
      2 2 2 R0
  2 ) 4 4 4
    - 4
      0 4
      - 4
        0 4
      -   4
          0
```

(9)
```
      4 5 8 R0
  2 ) 9 1 6
    - 8
      1 1
    - 1 0
        1 6
      - 1 6
          0
```

Page 40, Item 1:

(1)
```
           5 3 R3
  1 7 ) 9 0 4
     - 8 5
         5 4
       - 5 1
           3
```

(2)
```
             2 R42
  7 6 ) 1 9 4
     - 1 5 2
           4 2
```

(3)
```
             3 R35
  8 7 ) 2 9 6
     - 2 6 1
           3 5
```

(4)
```
           1 1 R70
  8 3 ) 9 8 3
     - 8 3
         1 5 3
       - 8 3
           7 0
```

Page 41, Item 1:

(1)
```
           1 1 R69
  7 7 ) 9 1 6
     - 7 7
         1 4 6
       - 7 7
           6 9
```

(2)
```
           1 1 R15
  3 7 ) 4 2 2
     - 3 7
         5 2
       - 3 7
           1 5
```

(3)
```
           1 0 R67
  8 2 ) 8 8 7
     - 8 2
         6 7
       -   0
           6 7
```

(4)
```
             1 R49
  5 1 ) 1 0 0
     -   5 1
           4 9
```

Page 42, Item 1:

(1)
```
           1 2 R9
  5 3 ) 6 4 5
     - 5 3
         1 1 5
       - 1 0 6
             9
```

(2)
```
             4 R34
  6 6 ) 2 9 8
     - 2 6 4
           3 4
```

(3)
```
           3 9 R5
  1 0 ) 3 9 5
     - 3 0
         9 5
       - 9 0
           5
```

(4)
```
             8 R29
  9 9 ) 8 2 1
     - 7 9 2
           2 9
```

124

Page 43, Item 1:

(1) 488 R8
13)632
−52
112
−104
8

(2) 2 R43
74)191
−148
43

(3) 15 R7
65)982
−65
332
−325
7

(4) 10 R71
81)881
−81
71
−0
71

Page 44, Item 1:

(1) 5 R48
67)383
−335
48

(2) 4 R8
38)160
−152
8

(3) 3 R22
91)295
−273
22

(4) 2 R36
94)224
−188
36

Page 45, Item 1:

(1) 10 R59
81)869
−81
59
−0
59

(2) 5 R13
83)428
−415
13

(3) 5 R4
22)114
−110
4

(4) 7 R20
68)496
−476
20

Page 46, Item 1:

(1) 10 R22
65)672
−65
22
−0
22

(2) 19 R7
30)577
−30
277
−270
7

(3) 3 R90
94)372
−282
90

(4) 10 R60
70)760
−70
60
−0
60

Page 47, Item 1:

(1) 35 R13
28)993
−84
153
−140
13

(2) 10 R14
26)274
−26
14
−0
14

(3) 34 R3
13)445
−39
55
−52
3

(4) 2 R71
95)261
−190
71

Page 48, Item 1:

(1) 6 R12
89)546
−534
12

(2) 69 R5
11)764
−66
104
−99
5

(3) 25 R30
32)830
−64
190
−160
30

(4) 5 R40
57)325
−285
40

Page 49, Item 1:

(1)
```
          4 R15
  6 0 ) 2 5 5
      - 2 4 0
          1 5
```

(2)
```
          6 R63
  7 0 ) 4 8 3
      - 4 2 0
          6 3
```

(3)
```
          5 R11
  2 6 ) 1 4 1
      - 1 3 0
          1 1
```

(4)
```
        1 4 R38
  4 1 ) 6 1 2
      -   4 1
        2 0 2
      - 1 6 4
          3 8
```

Page 50, Item 1:

(1)
```
          2 R89
  9 3 ) 2 7 5
      - 1 8 6
          8 9
```

(2)
```
          1 R67
  8 5 ) 1 5 2
      -   8 5
          6 7
```

(3)
```
        1 8 R8
  3 0 ) 5 4 8
      - 3 0
        2 4 8
      - 2 4 0
            8
```

(4)
```
        1 1 R9
  2 2 ) 2 5 1
      - 2 2
          3 1
        - 2 2
            9
```

Page 51, Item 1:

(1)
```
        1 5 R12
  6 1 ) 9 2 7
      - 6 1
        3 1 7
      - 3 0 5
          1 2
```

(2)
```
        1 1 R22
  4 8 ) 5 5 0
      - 4 8
          7 0
        - 4 8
          2 2
```

(3)
```
        1 9 R40
  4 8 ) 9 5 2
      - 4 8
        4 7 2
      - 4 3 2
          4 0
```

(4)
```
          5 R59
  6 4 ) 3 7 9
      - 3 2 0
          5 9
```

Page 52, Item 1:

(1)
```
          4 R68
  7 5 ) 3 6 8
      - 3 0 0
          6 8
```

(2)
```
        1 3 R7
  5 0 ) 6 5 7
      -   5 0
        1 5 7
      - 1 5 0
            7
```

(3)
```
        1 9 R24
  4 4 ) 8 6 0
      -   4 4
        4 2 0
      - 3 9 6
          2 4
```

(4)
```
          6 R45
  8 3 ) 5 4 3
      - 4 9 8
          4 5
```

Page 53, Item 1:

(1)
```
        1 3 R58
  6 5 ) 9 0 3
      - 6 5
        2 5 3
      - 1 9 5
          5 8
```

(2)
```
        4 8 R2
  1 1 ) 5 3 0
      - 4 4
          9 0
        - 8 8
            2
```

(3)
```
          1 R37
  9 5 ) 1 3 2
      -   9 5
          3 7
```

(4)
```
        1 0 R33
  8 7 ) 9 0 3
      - 8 7
          3 3
        -   0
          3 3
```

Page 54, Item 1:

(1)
```
        2 9 R1
  2 9 ) 8 4 2
      - 5 8
        2 6 2
      - 2 6 1
            1
```

(2)
```
        1 0 R18
  7 1 ) 7 2 8
      - 7 1
          1 8
        -   0
          1 8
```

(3)
```
        1 7 R3
  1 5 ) 2 5 8
      - 1 5
        1 0 8
      - 1 0 5
            3
```

(4)
```
          2 R9
  9 9 ) 2 0 7
      - 1 9 8
            9
```

Page 56, Item 1:
(1)5/6 (2)33/45 (3)5/6 (4)16/18 (5)9/12
(6)6/10 (7)23/18 (8)115/90 (9)17/14
(10)17/35 (11)57/45 (12)6/10
(13)101/72 (14)13/30 (15)21/20
(16)47/40 (17)15/12 (18)14/20
(19)31/30 (20)7/8 (21)13/10 (22)34/35
(23)23/40 (24)9/6 (25)132/90
(26)43/30 (27)41/40

Page 57, Item 1:
(1)3/4 (2)83/63 (3)10/21 (4)8/15
(5)17/36 (6)52/40 (7)30/36 (8)22/30
(9)33/45 (10)4/6 (11)15/12 (12)106/70
(13)11/9 (14)31/35 (15)26/30
(16)52/70 (17)35/72 (18)51/63
(19)35/72 (20)29/21 (21)13/12
(22)23/30 (23)26/20 (24)19/24
(25)14/10 (26)5/6 (27)56/45

Page 58, Item 1:
(1)43/30 (2)17/24 (3)26/40 (4)11/9
(5)22/30 (6)5/6 (7)50/63 (8)14/10
(9)65/90 (10)106/70 (11)17/14
(12)12/10 (13)11/15 (14)5/4 (15)68/63
(16)79/72 (17)11/9 (18)7/6 (19)25/18
(20)41/72 (21)66/63 (22)41/40
(23)39/42 (24)7/6 (25)7/6 (26)35/36
(27)9/8

Page 59, Item 1:
(1)19/18 (2)13/12 (3)5/6 (4)39/42
(5)25/24 (6)23/18 (7)71/45 (8)7/6
(9)10/8 (10)27/24 (11)3/6 (12)46/35
(13)42/40 (14)11/10 (15)5/6 (16)68/72
(17)27/35 (18)49/90 (19)12/8
(20)18/20 (21)6/10 (22)52/30
(23)37/70 (24)13/18 (25)50/36
(26)7/12 (27)55/40

Page 60, Item 1:
(1)41/28 (2)19/15 (3)13/12 (4)28/40
(5)11/12 (6)21/30 (7)7/10 (8)15/28
(9)31/20 (10)34/36 (11)69/56
(12)58/45 (13)32/30 (14)58/42
(15)22/24 (16)13/9 (17)32/30
(18)31/30 (19)43/45 (20)57/40
(21)31/24 (22)123/70 (23)29/36
(24)46/45 (25)26/24 (26)28/40
(27)97/56

Page 61, Item 1:
(1)50/40 (2)17/15 (3)11/10 (4)11/10
(5)7/8 (6)13/14 (7)78/63 (8)6/8
(9)49/40 (10)44/35 (11)32/30
(12)50/72 (13)13/12 (14)10/12
(15)27/30 (16)7/10 (17)74/72
(18)20/24 (19)19/12 (20)17/21
(21)16/21 (22)11/12 (23)82/70
(24)19/14 (25)38/30 (26)19/35
(27)13/18

Page 62, Item 1:
(1)2 18/40 (2)9 1/10 (3)2 24/40
(4)9 8/15 (5)5 18/28 (6)5 25/42
(7)2 14/30 (8)4 8/12 (9)3 33/72
(10)3 65/70 (11)3 3/28 (12)2 84/90
(13)4 43/63 (14)5 4/6 (15)8 6/8
(16)4 7/72 (17)7 3/20 (18)6 5/8

Page 63, Item 1:
(1)3 6/8 (2)3 56/90 (3)5 1/30 (4)6 7/8
(5)6 14/15 (6)6 22/36 (7)3 6/10
(8)4 46/70 (9)3 29/90 (10)4 39/45
(11)4 3/4 (12)2 19/20 (13)4 12/35
(14)3 37/90 (15)3 17/30 (16)9 1/6
(17)4 44/72 (18)10 5/6

Page 64, Item 1:
(1)3 10/12 (2)2 68/70 (3)3 44/72
(4)6 4/21 (5)5 1/6 (6)4 4/8 (7)3 5/9
(8)5 7/12 (9)7 3/12 (10)13 1/4
(11)5 1/20 (12)11 3/14 (13)4 35/72
(14)2 9/40 (15)6 10/12 (16)2 15/18
(17)3 7/24 (18)3 17/18

Page 65, Item 1:
(1)3 8/10 (2)4 1/63 (3)3 19/40
(4)2 8/10 (5)3 43/72 (6)6 3/6
(7)2 36/40 (8)3 2/18 (9)2 76/90
(10)3 1/24 (11)4 29/30 (12)3 7/9
(13)3 39/40 (14)5 16/20 (15)3 19/30
(16)3 43/70 (17)9 9/10 (18)3 3/42

Page 66, Item 1:
(1)5 7/20 (2)3 2/63 (3)4 5/6 (4)3 8/30
(5)2 76/90 (6)2 29/40 (7)3 22/56
(8)11 5/18 (9)3 4/40 (10)3 2/10
(11)3 24/40 (12)2 78/90 (13)6 2/36
(14)2 69/72 (15)3 10/42 (16)5 1/30
(17)5 1/28 (18)9 11/12

Page 67, Item 1:
(1)2 16/20 (2)3 58/63 (3)9 15/18
(4)6 2/9 (5)3 19/45 (6)5 32/35
(7)5 12/20 (8)3 15/63 (9)5 3/6
(10)3 13/42 (11)4 59/63 (12)8 20/24
(13)3 20/24 (14)6 3/14 (15)3 1/36
(16)5 27/40 (17)4 2/40 (18)3 38/63

Page 68, Item 1:
(1)3 27/40 (2)3 3/70 (3)3 13/30
(4)5 5/24 (5)3 5/42 (6)7 1/12
(7)3 33/42 (8)3 30/40 (9)4 12/72
(10)8 13/18 (11)3 11/28 (12)2 9/40
(13)5 23/35 (14)9 7/10 (15)3 37/40
(16)4 18/20 (17)4 11/20 (18)3 17/35

Page 69, Item 1:
(1)4 17/24 (2)4 4/18 (3)3 22/28
(4)4 13/18 (5)3 1/40 (6)7 4/21
(7)3 27/70 (8)5 26/45 (9)6 26/36
(10)2 22/36 (11)5 8/10 (12)9 10/12
(13)2 17/18 (14)5 16/30 (15)4 4/35
(16)7 4/15 (17)8 1/21 (18)2 66/70

Page 70, Item 1:
(1)4 2/40 (2)4 39/70 (3)4 12/72
(4)2 34/40 (5)4 11/30 (6)3 24/45
(7)2 9/40 (8)4 31/45 (9)9 5/6
(10)3 44/70 (11)2 7/8 (12)3 16/90
(13)9 7/10 (14)6 18/20 (15)2 22/35
(16)6 17/24 (17)3 13/14 (18)6 10/28

Page 71, Item 1:
(1)2/6 (2)2/8 (3)1/30 (4)1/10 (5)13/40
(6)9/42 (7)3/42 (8)2/8 (9)5/18
(10)17/30 (11)1/6 (12)2/12 (13)14/24
(14)1/18 (15)3/36 (16)32/90 (17)1/40
(18)3/28 (19)11/45 (20)7/15 (21)8/30
(22)27/40 (23)26/90 (24)11/24
(25)7/12 (26)6/40 (27)1/4

Page 72, Item 1:
(1)7/72 (2)3/14 (3)4/21 (4)4/8 (5)1/18
(6)24/63 (7)1/20 (8)1/12 (9)1/9 (10)1/8
(11)7/20 (12)9/40 (13)1/4 (14)7/20
(15)2/10 (16)2/8 (17)4/63 (18)13/56
(19)7/30 (20)2/30 (21)1/4 (22)8/90
(23)4/40 (24)1/6 (25)19/36 (26)16/42
(27)19/72

Page 73, Item 1:
(1)17/45 (2)15/42 (3)6/36 (4)2/8 (5)1/8
(6)26/45 (7)22/42 (8)7/36 (9)3/8
(10)1/36 (11)2/12 (12)7/24 (13)1/10
(14)33/56 (15)2/18 (16)2/15 (17)11/42
(18)3/36 (19)4/15 (20)16/70 (21)3/6
(22)2/24 (23)3/36 (24)5/18 (25)44/90
(26)1/6 (27)9/30

Page 74, Item 1:
(1)19/30 (2)1/10 (3)3/8 (4)5/14 (5)5/30
(6)2/6 (7)5/12 (8)3/10 (9)5/14 (10)1/9
(11)13/20 (12)3/10 (13)3/10 (14)6/45
(15)1/6 (16)29/42 (17)3/12 (18)7/30
(19)41/56 (20)1/10 (21)5/20 (22)6/20
(23)5/8 (24)1/6 (25)7/20 (26)22/90
(27)13/28

Page 75, Item 1:
(1)3/10 (2)1/4 (3)1/4 (4)1/6 (5)2/21
(6)1/20 (7)6/10 (8)3/8 (9)4/70 (10)1/72
(11)1/12 (12)9/30 (13)7/10 (14)2/36
(15)5/9 (16)11/20 (17)37/72 (18)5/12
(19)5/14 (20)1/12 (21)11/21 (22)1/8
(23)19/30 (24)5/9 (25)1/14 (26)3/30
(27)1/6

Page 76, Item 1:
(1)1/4 (2)11/35 (3)1/4 (4)5/12 (5)2/6
(6)3/6 (7)3/10 (8)4/15 (9)11/36
(10)2/21 (11)7/90 (12)1/4 (13)1/14
(14)7/12 (15)4/10 (16)1/40 (17)4/15
(18)2/70 (19)4/15 (20)11/24 (21)7/20
(22)5/30 (23)5/14 (24)6/63 (25)3/8
(26)10/56 (27)2/6

Page 77, Item 1:
(1)2 2/24 (2)34/63 (3)1 11/40 (4)3 3/8
(5)2 33/36 (6)1 13/21 (7)13/20 (8)7/20
(9)1/63 (10)1 11/36 (11)23/30
(12)1 3/20 (13)1 3/18 (14)2 8/20
(15)19/70 (16)1/40 (17)2/42 (18)1 2/20

Page 78, Item 1:
(1)3 3/9 (2)8/18 (3)14/36 (4)2 17/24
(5)1 6/8 (6)2/12 (7)2 4/15 (8)1 4/10

(9)4 11/21 (10)1 9/35 (11)22/24
(12)2 1/20 (13)6 5/8 (14)1 2/18
(15)6/90 (16)6 2/10 (17)13/14
(18)11/35

Page 79, Item 1:
(1)22/35 (2)1 8/15 (3)35/36 (4)53/56
(5)38/63 (6)5 8/30 (7)1 7/18 (8)31/90
(9)2 25/36 (10)8/30 (11)20/56
(12)1 1/8 (13)1 9/20 (14)1 9/40
(15)7 2/10 (16)9/56 (17)3/14 (18)24/40

Page 80, Item 1:
(1)6 9/14 (2)2 19/36 (3)1 3/4 (4)17/24
(5)1 14/18 (6)36/70 (7)2/9 (8)1 33/56
(9)5/72 (10)51/63 (11)1 2/40
(12)1 22/30 (13)11/30 (14)2 5/6
(15)1 1/30 (16)4 2/8 (17)37/63
(18)1 1/42

Page 81, Item 1:
(1)1/8 (2)38/72 (3)2 3/36 (4)4 4/30
(5)1 13/30 (6)1 7/20 (7)17/56 (8)5/6
(9)12/56 (10)1 11/15 (11)6/40
(12)4 1/4 (13)47/63 (14)8 3/10
(15)3 22/30 (16)1 2/42 (17)38/40
(18)45/70

Page 82, Item 1:
(1)31/42 (2)2 1/15 (3)1 1/18 (4)6 4/6
(5)8/63 (6)52/72 (7)13/63 (8)1 27/35
(9)3 14/15 (10)10/36 (11)18/20
(12)1 2/10 (13)2 15/36 (14)3 1/6
(15)1 4/30 (16)6/40 (17)1 18/70
(18)3/4

Page 83, Item 1:
(1)1 2/15 (2)1 2/30 (3)19/72 (4)29/70
(5)1 5/6 (6)3 4/30 (7)3 2/21 (8)1 38/63
(9)1 1/10 (10)19/72 (11)17/20
(12)1 11/12 (13)2/24 (14)2 1/36
(15)1 19/20 (16)62/90 (17)2/40
(18)12/90

Page 84, Item 1:
(1)22/42 (2)2 7/12 (3)15/90 (4)4 7/30
(5)2 5/6 (6)1 5/8 (7)2 3/10 (8)27/28
(9)26/40 (10)4 20/24 (11)3 15/36
(12)3 1/12 (13)1 4/9 (14)4 7/24
(15)8/30 (16)2 7/15 (17)16/45
(18)2 1/20

Page 85, Item 1:
(1)18/40 (2)2 1/4 (3)1 9/10 (4)17/20
(5)19/40 (6)3 5/6 (7)20/24 (8)1 1/36
(9)19/30 (10)2 2/35 (11)3 1/10
(12)1 2/6 (13)3 2/6 (14)2 2/6 (15)2 5/9
(16)5 3/8 (17)2 4/20 (18)37/72

Page 87, Item 1:
(1)14.106 (2)8.030 (3)3.781 (4)4.652
(5)7.909 (6)13.156 (7)7.597 (8)11.345
(9)8.184 (10)16.883 (11)13.991
(12)10.630 (13)16.015 (14)6.727
(15)12.807 (16)5.490 (17)12.666
(18)7.590 (19)4.673 (20)17.985
(21)8.026 (22)13.966 (23)11.476
(24)4.642 (25)3.483 (26)8.761
(27)2.065 (28)9.373

Page 88, Item 1:
(1)13.881 (2)13.260 (3)13.003 (4)6.297
(5)7.900 (6)3.227 (7)10.349 (8)12.137
(9)9.660 (10)7.083 (11)13.711
(12)8.356 (13)8.728 (14)8.661

(15)8.181 (16)5.156 (17)9.648
(18)12.703 (19)6.688 (20)11.940
(21)9.214 (22)13.353 (23)5.783
(24)15.144 (25)1.889 (26)16.239
(27)9.264 (28)10.962

Page 89, Item 1:
(1)16.516 (2)9.383 (3)9.202 (4)10.797
(5)13.061 (6)10.276 (7)13.204 (8)9.180
(9)9.474 (10)4.950 (11)15.182
(12)13.133 (13)9.762 (14)4.065
(15)11.545 (16)9.802 (17)12.068
(18)5.004 (19)10.816 (20)11.112
(21)4.832 (22)11.477 (23)16.421
(24)9.259 (25)12.136 (26)13.325
(27)8.974 (28)8.601

Page 90, Item 1:
(1)15.579 (2)15.655 (3)7.169 (4)7.583
(5)5.201 (6)5.685 (7)3.514 (8)10.843
(9)9.025 (10)9.388 (11)4.984 (12)8.842
(13)10.108 (14)7.448 (15)16.193
(16)8.442 (17)5.965 (18)14.005
(19)13.290 (20)7.820 (21)6.606
(22)17.145 (23)10.564 (24)11.600
(25)15.082 (26)3.485 (27)11.418
(28)3.464

Page 91, Item 1:
(1)10.858 (2)12.471 (3)3.809 (4)6.829
(5)12.136 (6)14.204 (7)10.421 (8)5.119
(9)11.648 (10)17.202 (11)10.490

(12)12.294 (13)15.414 (14)16.566
(15)14.382 (16)13.040 (17)14.686
(18)3.665 (19)8.982 (20)5.205
(21)5.649 (22)1.681 (23)9.503
(24)7.423 (25)11.257 (26)8.272
(27)14.652 (28)11.647

Page 92, Item 1:
(1)11.914 (2)5.394 (3)14.145 (4)14.431
(5)16.687 (6)8.717 (7)12.224 (8)11.084
(9)5.202 (10)8.619 (11)8.952 (12)3.164
(13)2.978 (14)3.601 (15)9.561
(16)12.708 (17)11.106 (18)9.513
(19)7.554 (20)8.252 (21)8.450
(22)10.165 (23)12.366 (24)3.189
(25)9.040 (26)16.409 (27)12.641
(28)8.175

Page 93, Item 1:
(1)1.997 (2)7.784 (3)0.858 (4)0.585
(5)4.555 (6)0.402 (7)4.406 (8)6.066
(9)2.285 (10)3.189 (11)0.723 (12)1.636
(13)1.871 (14)4.698 (15)1.173
(16)4.712 (17)9.784 (18)5.385
(19)3.775 (20)2.361 (21)5.309
(22)2.452 (23)1.557 (24)4.785
(25)0.989 (26)3.989 (27)5.354
(28)4.978

Page 94, Item 1:
(1)7.183 (2)1.678 (3)2.850 (4)0.272
(5)0.310 (6)3.234 (7)3.481 (8)8.528
(9)0.118 (10)8.876 (11)5.416 (12)3.619
(13)1.575 (14)8.164 (15)2.742
(16)2.960 (17)0.812 (18)4.081
(19)7.634 (20)5.678 (21)7.236
(22)4.562 (23)0.403 (24)0.689
(25)0.794 (26)6.119 (27)6.511
(28)3.600

Page 95, Item 1:
(1)1.207 (2)0.338 (3)2.292 (4)7.138
(5)3.821 (6)2.591 (7)7.478 (8)2.382
(9)6.598 (10)2.220 (11)1.777 (12)4.197
(13)2.107 (14)5.102 (15)2.689
(16)6.097 (17)5.926 (18)1.272
(19)0.453 (20)0.747 (21)3.345
(22)9.262 (23)5.185 (24)0.504
(25)6.738 (26)2.479 (27)1.486
(28)8.717

Page 96, Item 1:
(1)1.492 (2)5.830 (3)3.156 (4)0.625
(5)5.542 (6)6.814 (7)0.127 (8)1.641
(9)5.916 (10)4.291 (11)4.341 (12)1.420
(13)6.253 (14)0.645 (15)6.439
(16)2.354 (17)5.922 (18)1.656
(19)6.306 (20)2.663 (21)2.764
(22)5.740 (23)5.197 (24)0.107
(25)4.382 (26)5.979 (27)1.477
(28)2.115

Page 97, Item 1:
(1)0.233 (2)4.158 (3)3.347 (4)3.933
(5)3.322 (6)5.189 (7)0.316 (8)0.096
(9)1.446 (10)4.735 (11)3.426 (12)3.043
(13)2.356 (14)7.267 (15)0.749
(16)4.822 (17)1.403 (18)5.099
(19)1.609 (20)8.532 (21)2.658
(22)0.059 (23)1.535 (24)2.879
(25)1.974 (26)1.190 (27)1.772
(28)1.046

Page 98, Item 1:
(1)7.486 (2)2.507 (3)1.939 (4)0.527
(5)1.490 (6)2.504 (7)1.825 (8)2.426
(9)3.222 (10)6.010 (11)0.618 (12)2.837
(13)3.859 (14)2.464 (15)4.666
(16)1.834 (17)2.774 (18)2.079
(19)5.374 (20)6.576 (21)6.436
(22)1.216 (23)9.080 (24)0.550
(25)1.709 (26)1.220 (27)5.687
(28)4.585

Page 99, Item 1:
(1)7.7 (2)39.56 (3)31.5 (4)10.79
(5)38.64 (6)3.6 (7)13.77 (8)33.93
(9)53.76 (10)34.86 (11)53.69 (12)7.48
(13)1.56 (14)60.48 (15)1.68 (16)8.1
(17)4.32 (18)25.96 (19)38.8 (20)1.14

Page 100, Item 1:
(1)44.18 (2)6.48 (3)17 (4)69.6 (5)6.48
(6)43.16 (7)88.35 (8)62.4 (9)20.88
(10)5.28 (11)0.51 (12)48.36 (13)12.22
(14)6.4 (15)41.16 (16)2.07 (17)1.68
(18)26.6 (19)14 (20)11.02

Page 101, Item 1:
(1)51 (2)26.73 (3)19.14 (4)23.32
(5)13.35 (6)14.22 (7)34.84 (8)39.95
(9)34.2 (10)73.6 (11)64.02 (12)1.9
(13)14.74 (14)27.72 (15)23.8 (16)37.84
(17)15.41 (18)37.62 (19)28.5 (20)5.13

Page 102, Item 1:
(1)1.59 (2)1.6 (3)1.02 (4)3.76 (5)49
(6)23.68 (7)12.35 (8)42.78 (9)2.28
(10)4.75 (11)5.16 (12)7.92 (13)22.79
(14)7.5 (15)16.34 (16)27.52 (17)5.5
(18)22.62 (19)4.8 (20)2.61

Page 103, Item 1:
(1)13.34 (2)10.89 (3)33.37 (4)73.15
(5)2.25 (6)2.72 (7)37.44 (8)16.8 (9)23
(10)21.93 (11)26.07 (12)13 (13)3.33
(14)15.05 (15)4.25 (16)68.8 (17)11.6
(18)25.96 (19)8.04 (20)84.39

Page 104, Item 1:
(1)23.65 (2)1.47 (3)58.41 (4)4.62
(5)0.12 (6)1.75 (7)62.4 (8)28.98 (9)42.9
(10)78.85 (11)2.72 (12)7.7 (13)10.73
(14)15.48 (15)14.49 (16)5.88 (17)92.07
(18)18.9 (19)78.85 (20)46.4